Table of Contents

FOREWORD

Alternative transportation fuels may enhance the energy security of countries that are Members of the International Energy Agency and may help meet goals for an improved environment. Oil consumed in transport today accounts for well over half of total final oil consumption in IEA countries, and the share is steadily growing. Indeed, oil use for transport is roughly equivalent to the total amount of oil which the IEA countries import. The potential for substitution of fuels based on natural gas or alcohol, as well as programmes that encourage oil to be used more efficiently, must thus be given priority consideration by IEA energy planners looking forward to the 1990s.

The use of petroleum-based transport fuels accounts for a significant fraction of air pollution, including emissions of sulfur dioxide, nitrogen oxides, volatile organic compounds, carbon monoxide, and carbon dioxide. Substitute fuels have the potential to reduce such emissions. For example, methanol or ethanol fuel could reduce the nitrous oxides, carbon monoxide, and particulates that contribute to smog in urban areas. Over a given distance travelled, compressed natural gas would emit roughly one-fifth less carbon dioxide than oil-based fuels, and alcohol fuels from biomass could have even lower net carbon dioxide emissions when the full cycle of fuel production and use is considered.

Yet despite their potential benefits for energy security and the environment, alternative transportation fuels are still surrounded by uncertainties as to their technical feasibility and cost. There may also be difficulties in bringing them to market, particularly related to distribution infrastructure and consumer acceptance. Question marks remain as to the sources from which greatly increased volumes of these alternatives would be acquired. Technological progress and market development will therefore be necessary before their potential benefits can be fully realized. The present study attempts to identify which transport fuel options are most promising for the next two decades and to highlight key technical obstacles to their use. Specific recommendations are offered as to research and development priorities, with an aim to facilitating the adoption of each fuel in a timely fashion.

The IEA Secretariat was guided in the preparation of this study by officials of Member governments and a group of experts drawn from government and industry. I am most grateful to all of them for their help, without which the study could not have been completed. The study's findings have been formally approved by the IEA Member countries through the Committee on Energy Research and Development, the Standing Group on Long-Term Co-operation, the Standing Group on the Oil Market, and the IEA Governing Board.

Helga Steeg
Executive Director
International Energy Agency

EXECUTIVE SUMMARY

Introduction

The transport sector is still almost entirely dependent on liquid fuels derived from oil, but the oil production of countries which are members of the International Energy Agency (IEA) is expected to decline during the 1990s. Accordingly, in meetings of 11th May 1987 and 30th May 1989, the energy Ministers of IEA Member countries decided that more attention should be focussed on efforts to promote diversification among these fuels. Thus the present study assesses the energy security, economic and environmental aspects of the main alternative or substitute fuel options which could contribute to such diversification. It then identifies the most significant issues which need to be addressed in the context of substitute fuels entering the market place. Prominent among these issues are R&D needs, including areas for international collaboration.

Scope of the Study

The time horizon for the study was placed at roughly the year 2005. Such a horizon is close enough to allow reasonably firm judgements, but may also allow sufficient time for some technological change to occur. Only road transport was considered, since this accounts for roughly four-fifths of the petroleum-based liquid fuels that IEA members use for transportation. Detailed analysis was then performed for those fuels that appeared to have significant potential to contribute to energy security in the chosen timeframe:

— fuels from very heavy oils (VHOs) and tar sands;
— natural gas, in compressed (CNG) or liquefied (LNG) form;

— methanol (in blends of 85 percent methanol content and above);
— ethanol (in blends of 85 percent ethanol content and above); and
— synthetic fuels from natural gas.

Liquefied petroleum gas (LPG), low-level alcohol blends, electric vehicles, hydrogen, synthetic fuels from coal or oil shale and vegetable oils were considered only briefly. Some of these were seen as being unlikely to make a significant contribution in the chosen time-frame. Others could only become commercial niche technologies, with narrowly defined markets.

The Economics of Transport Fuel Options

The economics of each alternative transport fuel depend upon both the cost of production and the additional cost of distribution and end-use. Production costs, in turn, depend upon the abundance or scarcity of the resources from which the fuel is produced, as well as the technology which is available to extract those resources which are available. The additional cost of distribution and end use depends on the extent to which means are already in place to deliver the fuel. This is important since gasoline and diesel fuel made from VHOs or natural gas require relatively minor changes to existing distribution and end-use systems, whereas CNG and the alcohol fuels require major modifications.

Since natural gas is a feedstock for several of the fuels that were analyzed, the question of its availability is considered in some detail. An important consideration is the extent to which energy security is enhanced by relying upon gas resources rather than crude oil. OECD countries have about 7 percent of world oil reserves and about 14 percent of world gas reserves. The development of gas markets during a time when there could be concurrent increasing demands for transport fuels, electric power, industrial and residential uses is an emerging issue.

The other resources considered are VHOs (including tar sands and bitumens), coal and biomass. Total resources of VHOs are estimated to at least equal those of conventional crude oils, but are much less well known; a large proportion of discovered VHOs are located in Canada. Coal resources are abundant in many countries, so their use has a clearer potential to enhance energy security. Biomass resources have the advantage of being renewable, and they may be abundant in certain areas, but large-scale fuel production from biomass faces a number of technical challenges and relatively high cost.

The table above shows the estimated cost ranges of alternative fuels based on current oil and gas prices and the technology now available. CNG and VHO products, at the lower end of their estimated cost ranges, may be economically competitive with conventional gasoline at present. Methanol and synthetic gasoline made from natural gas may be close to competitive, under optimistic assumptions about gas prices. Methanol from coal or biomass and ethanol from biomass have a cost at least double that of gasoline at current oil prices and with current technology.

Comparative Costs of Substitute Fuels Today

Fuel	Overall Cost (1987 US dollars per barrel-gasoline energy equivalent)
Crude Oil (assumed price)	$ 18
Conventional Gasoline	$ 27
CNG	$ 20-46
VHO Products	$ 21-34
Methanol (from gas)	$ 30-67
Synthetic Gasoline (from gas)	$ 43-61
Diesel (from gas)	$ 69
Methanol (from coal)	$ 63-109
Methanol (from biomass)	$ 64-126
Ethanol (from biomass)	$ 66-101

A conversion to alternative fuels would require significant investment in new production and distribution equipment. There could be significant problems associated with applying unfamiliar technologies on a large scale in an uncertain market environment. Hence, before such a major commitment of resources takes place, oil prices may have to rise significantly or technological development may have to render the alternatives less costly.

A further important issue is the relationship between the prices of conventional oil, natural gas and the alternative fuels. If a given country or company produces both oil and gas, it is apt to price its gas so as to maximise its overall return from the two resources. This may, in turn, affect the market price of methanol from natural gas. The extent to which production of alternative fuels such as methanol can provide a limitation on world oil prices is thus unclear.

Indications are that the cost of most alternative fuel options could be reduced substantially through the development of new and improved technologies. This is particularly true of the more advanced fuel production processes, which are naturally those that are least developed. Since most alternative fuels are still far more costly than conventional products, the technical opportunities must be seized if such fuels are to play a major role. Given the potential of alternative fuels to enhance energy security by reducing reliance on oil, the higher costs of these fuels present an opportunity for appropriately targeted research, development and demonstration.

Vehicle End-Use Considerations

Since a major goal of fuel diversification is to reduce dependence on oil, it was necessary to consider the extent to which reduced oil use might be achieved through increased fuel efficiency as an alternative or complementary approach to the use of new fuels. Energy efficiency improvements enhance both energy security and protection of the environment. Yet in many cases, they do not require major changes to the distribution and end-use infrastructure.

It appears that transport fuel consumption could continue to rise in most IEA member countries, despite continuing improvements in the efficiency of individual vehicles. Indeed, there is evidence that the fleet average efficiency in certain countries may decline for a period as more affluent consumers buy larger vehicles. Hence, there may be substantial opportunities to reduce oil reliance through measures to promote increased vehicle efficiency. But given the robustness of fuel demand, it appears that programmes to increase efficiency would complement rather than replace the development of alternative fuels.

The increased use of very heavy oils could possibly have a modest negative impact on vehicle end-use efficiency in the absence of improved technology for their use. Engine performance with them is generally comparable to that with conventional fuels. Products from VHOs are similar to those from conventional oil but may have reduced quality. There can be problems if fuels are not sufficiently upgraded.

Conversion of vehicles to compressed natural gas may tend to reduce fuel efficiency unless there is further development of engines that are specifically designed for its use. CNG is already used quite extensively in certain countries and regional contexts. But at present, CNG performance in Otto engines modified by conversion kits (that is to say, modified spark-ignition engines) is slightly below that for gasoline in unmodified engines. The capital cost of conversion is significant, the gas cylinders reduce efficiency by increasing engine weight, and less fuel can typically be stored in the vehicle. Research and development is being targeted at reducing the weight of gas cylinders through the use of new materials. However, it appears that CNG is quite suitable for use in Diesel engines as well. This may point to larger, higher-mileage Diesel vehicles, for which the proportionate increase in weight is smaller. In addition, industry is beginning to design engines specifically for use with CNG, and this should lead to greater efficiency and lower cost compared to retrofits.

Use of liquefied natural gas, particularly for large off-road vehicles, is also being studied. Typically, LNG is stored in cryogenic tanks and is vaporised as required for use in the engine cylinders. Since the cryogenic storage tanks are bulky, it may prove impractical to develop the use of LNG in smaller vehicles. This would not preclude the adoption of LNG in larger transportation systems, particularly in rail systems, where fuel might be tendered to vehicles without being stored in them.

Although they have lower energy value per unit volume than does gasoline, the alcohol fuels have better combustion characteristics for use in Otto engines. The recent development of fuel-flexible vehicles, able to run on any mixture of alcohol and gasoline, is significant for aiding a potential transition to dedicated alcohol-fuelled vehicles. Fuel-flexible vehicles are at the fleet trial stage and appear to be operating satisfactorily, though certain issues remain to be resolved. These include cold climate driveability, fuel sensor reliability, engine oil specifications, and emission certification and compliance.

The properties of alcohol fuels are not immediately favourable to improved efficiency of the Diesel cycle, but results comparable to those with conventional diesel fuel have now been obtained. The technology is less well developed than for Otto engines. A number of manufacturers are pursuing different approaches, and it is not yet clear whether a single preferred alcohol/Diesel technology will emerge. A number of small-scale fleet trials are underway.

The "methanol engine", designed from the outset to make full use of the properties of alcohols, is still only a concept, but one on which R&D activities should be focussed. In general, the use of methanol engines might well await the establishment of an infrastructure for methanol delivery. However, they could be introduced beforehand in centrally fuelled fleets.

Environmental Characteristics of Substitute Fuels

In addition to energy security considerations, environmental issues may provide a large part of the impetus for introduction of alternative fuels in some IEA Member countries. But to justify their introduction on environmental grounds, alternatives must perform better than existing fuels.

Particulate emissions from vehicles are a major factor to consider in evaluating the relative environmental merits of alternative fuels. Diesel fuel from VHOs, unless sufficiently upgraded, has higher particulate emissions (smoke) than does petroleum-based diesel fuel, CNG or alcohol fuel. The United States and Canada are bringing in more stringent regulations to limit particulate emissions. Sweden and Switzerland also have firm plans on particulates. If environmental concerns persist, other IEA countries may eventually follow suit. Hence, possible future regulation of particulate emissions should be borne in mind when designing research programmes.

Combustion of natural gas in vehicles results in reduced hydrocarbon and carbon monoxide emissions, but raises some questions regarding levels of nitrogen oxides. Relatively little study has been undertaken in this area, and more work is needed on combustion and suitable catalysts. It seems that there may be potential for cost savings through reduced need for emission controls in gas-fuelled vehicles.

More test results are available for the alcohol fuels, although it is important to note that these are usually for engines optimised for fuel economy rather than emissions. In principle, emissions from alcohol fuels should be lower than those from gasoline or diesel fuel. Results available to date show that alcohol fuels may reduce ozone emissions in Otto cycle engines because the exhaust and evaporative emissions are less reactive, but this remains to be confirmed. Particulate emissions from compression/ignition engines may also be greatly reduced through use of methanol. On the other hand, much research and development remains to be done if methanol is to fulfill its theoretical potential for emissions reductions at a reasonable cost. It may be possible to reduce emission control costs for alcohol fuels below those for conventional fuels. However, the question of formaldehyde control will first need to be resolved, either by ad-hoc tuning of engines or by the use of a simple oxidation catalyst.

The question of impact on global climate change is only briefly addressed in this study, although the general issue of "greenhouse" gases has achieved great prominence. Combustion of fuels from biomass would theoretically yield zero net production of greenhouse gases, provided that replacement biomass crops were planted to maintain the methanol economy. After biomass fuels, burning of CNG and methanol from gas would have the next lowest greenhouse emissions. Both biomass- and gas-based fuels would appear to be more environmentally benign than petroleum-based fuels. However, significant uncertainties remain about the relative merits of alternative fuel cycles and their environmental impacts. Before the use of any fuel is expanded on environmental grounds, it would be helpful to understand these impacts better.

Marketing Aspects

Since the present study is a technology assessment, marketing aspects are simply identified rather than discussed in detail. It is, nevertheless, essential to recognise them when designing technological programmes. For example, a typical marketing difficulty is the familiar circular dilemma: no-one will produce special vehicles if there is no alternative fuel available, and no-one will produce an alternative fuel if there are no vehicles available to use it. This marketing problem may be overcome through research and development of fuel-flexible vehicles in which either alternative or conventional fuel may be used.

The potential obstacles to marketing alternative fuels are numerous. On the supply side, these include fuel price, availability, standardisation and safety. At the point of end-use, they include vehicle price, performance, optimisation and safety. Environmental regulations, not aimed at promoting the role of alternative fuels per se, could also play a major role. Given the number and complexity of the obstacles to bringing new fuels to market, research and development efforts should be focused on alternatives which have not only technical merit, but also the potential to surmount these obstacles.

Chapter I

SCOPE OF THE STUDY

The 11th May 1987 meeting of IEA Ministers recommended that "more attention should be focused on R&D efforts to promote a higher degree of diversification" in transportation fuels. Within the transportation sector, Energy Ministers found that technology had been slow to provide cost-effective alternatives to liquid fuels based on petroleum. This view was reiterated by the Energy Ministers on 30th May 1989. Although oil use had been cut substantially through efficiency gains, there remained an opportunity to further reduce oil use through development of substitute fuels.

The primary purpose of this study is to assess the potential of alternative transportation technologies to contribute to energy security. Since domestic oil production in several IEA Member countries is expected to decline substantially in the 1990s, such an assessment would seem to be timely. Energy security does not imply national self-sufficiency, but rather the availability of energy in physical quantities and at cost levels which help sustain economic growth and prosperity.

The study may be characterised as a technology assessment. It looks at the prospects for diversification of transport fuels in terms of the full range of production and use factors which could influence the commercialisation of alternative fuels not widely adopted at present. The study was conducted by the IEA Secretariat, under the guidance of the IEA's Committee on Energy Research and Development, with the help of an ad-hoc group of experts. It addresses what is known and what needs to be known about alternative or substitute transport fuel technologies. It also seeks to identify what opportunities exist for international collaboration in this field. A good deal of attention is given to the environmental implications of the most promising alternative fuel options. Environmental considerations have provided much of the recent impetus for the development of substitute fuels. Yet the best technology choices from an environmental perspective will not necessarily form the best strategy from an energy security point of view.

It is important to emphasize from the outset that only the most important issues are outlined, backed as far as possible with quantitative data from the available literature. No definitive answers are provided to all the complex questions involved. However, the study may lead to more detailed examination of the issues it identifies as worthy of follow-up work.

The time-frame for the study extends through the year 2005. This time-frame was chosen to allow consideration of two major possibilities: tightening of oil markets and introduction of new technologies. The possible impact of tighter oil markets is assessed by examining two crude oil price scenarios (both expressed in 1987 U.S. dollars): prices rising gradually from $17.50 per barrel to $30 per barrel by 2000 but stable thereafter; and prices remaining constant at $17.50 per barrel. With regard to technologies that are now being researched or demonstrated, the chosen time frame is long enough to allow for their possible use in the vehicle fleet, yet short enough to constrain the analysis to technologies which already show significant promise. Beyond this time frame, the degree of uncertainty may simply be too great to allow a reasonable assessment.

Table I-1: **Final Energy Consumption in OECD Countries, by Sector, 1987**

Category	Total Consumption (Million tons oil equivalent)		Consumption Share (percent of total)	
Transport		829.3		30.2
— Road	682.9		82.4	
— Air	105.4		12.7	
— Other	41.0		4.9	
TRANSPORT TOTAL	829.3		100.0	
Industry		912.2		33.3
Residential/Commercial and Other Sectors		898.3		32.8
Non-energy Use		102.4		3.7
GRAND TOTAL		2 742.2		100.0

Source: IEA/OECD; 1989 (Ref. 7).

As Tables I-1 and I-2 show, road transport accounts for by far the largest single component of both total energy and total oil consumption in OECD Member countries. In 1987, the transport sector was responsible for 31 percent of total energy and 56 percent of total oil consumption. Road transport accounted for 82.4 percent of total transport energy consumption and 82.9 percent of oil consumption in the transport sector. A breakdown of current transport energy use by fuel type is provided in Table I-3.

Table I-2: **Final Consumption of Oil in OECD Countries, by Sector, 1987**

Category		Total Consumption (Million tons oil equivalent)		Consumption Share (percent of total)
Transport		823.0		56.5
— Road	682.4		82.9	
— Air	105.3		12.8	
— Other	35.3		4.3	
TRANSPORT TOTAL	823.0		100.0	
Industry		271.0		18.6
Residential/Commercial and Other Sectors		260.0		17.9
Non-energy Use		102.5		7.0
GRAND TOTAL		1 456.5		100.0

Source: IEA/OECD; 1989 (Ref. 7).

Table I-3: **Final Energy Consumption in Road Transport in OECD Countries, by Fuel, 1987**

Fuel Type	Mtoe	Percent
Gasoline	507.8	74.3
Automotive Diesel Oil	168.6	24.7
LPG	6.0	0.9
Natural Gas	0.5	0.1
Alcohols	0.0	0.0
TOTAL	682.9	100.0

Source: IEA/OECD; 1989 (Ref. 10).

Since road transport is such a major component of total transportation, the study deals with road transport exclusively. Furthermore, it focusses only on those fuels that already appear to have the potential to contribute to energy security. "Significant contribution" is used to mean that the fuel could potentially substitute for oil in at least a whole sub-sector, such as heavy vehicles, and that penetration up to one-third of the market may be possible within the 2005 time horizon of the study. Based on these screening criteria, the study includes assessments of :

— fuels from very heavy oils and tar sands;
— natural gas for vehicles, in compressed or liquefied form;
— methanol (in blends of 85 percent methanol content and above);
— ethanol (in blends of 85 percent ethanol content and above);
— synthetic fuels from natural gas.

The likely actual contribution from each of these fuels is discussed in Chapter III.

A number of fuels were not considered further. Brief details of the relevant reasons are as follows:

— Liquefied petroleum gas (LPG) is largely a by-product of crude oil refining and is therefore classed as a petroleum-based fuel, and one which moreover could make only a limited contribution to diversification. In addition, it is in widespread commercial use, with little need for technological development. Where technological advances might be useful, for example in Diesel engines, much of the discussion under compressed natural gas (CNG) is relevant.

— The category of other oxygenated fuels (i.e. apart from methanol and ethanol), includes ethers, particularly methyl-tertiary-butyl ether, and higher alcohols such as iso-propanol and tertiary-butanol. These are widely used in low-level blends to replace lead as an octane improver. These blends are fully compatible with the existing system and do not require separate pump labelling. Such fuels are considered as gasoline octane boosters as well as gasoline extenders; they are currently obtained from chemical by-products, refinery streams and field butanes. Their production and use as gasoline components constitute fully commercial technologies.

— Low-level alcohol blends can also, by definition, make only a limited contribution to fuel diversification. The choice of suitable materials to withstand long-term corrosion and swelling appears to be the only major end-use problem. Production aspects relating to methanol and ethanol are in any case considered in detail in this study.

— Electric vehicles are attractive from both energy security and environmental viewpoints in that they would move away from liquid or gaseous fuels entirely, but batteries have yet to overcome the penalty of their very low energy density. For example, gasoline has 37 MJ/kg, diesel fuel 36 MJ/kg, methanol 18 MJ/kg and the lead-acid battery 0.08 MJ/kg. Large R&D programmes aiming at up to threefold increases in energy density have been undertaken for different battery types, and some more modest programmes continue. Large-scale market penetration in the time-frame of this study is unlikely, although niche markets may slowly increase, for example, for public service vehicles with limited itineraries. Fuel cells have considerable potential for stationary applications, but their weight would seem to limit their initial transport applications to large units such as locomotives and possibly ships.

— Hydrogen produced from water, via electrolysis or pyrolysis, has clear advantages from supply and environmental points of view. But production by electrolysis is still about five times more costly than by steam reforming of methane, though research continues to reduce this cost. Small-scale vehicle trials indicate that there are no major problems with vehicle operation although storage systems are still too bulky and heavy for practical use. Major modifications would need to be made to the distribution system. Again, large scale deployment for road vehicles will be beyond the time frame of this study. International collaboration promoted by the IEA has covered experimental work on thermochemical, electrolytic and photocatalytic production, as well as studies on hydrogen storage, conversion, safety and future market prospects.

— Synthetic fuels from coal or oil shale have a large resource base and can undoubtedly be produced, but seem unlikely to become economically competitive with other alternative fuels within the study's time horizon. A minor exception could be the exploitation of some particularly rich shale deposits. However, methanol produced from gas is included since it could provide a transition to methanol from coal later on.

— As vegetable oils are derived from biomass, the limited availability of land is likely to constrain their potential to provide for large-scale fuel substitution in IEA Member countries. Furthermore, vegetable oils can only be used in Diesel engines or other specially designed engines, and they generally require esterification to over come viscosity problems. Small scale R&D programmes continue in some countries, concentrating on genetic improvement of plant species. There could be some potential for use in agricultural equipment, particularly in developing countries, and especially if they can be produced from simple cold pressing and inexpensive processing.

Chapter II

FEEDSTOCKS AND THE ECONOMICS OF TRANSPORT FUEL OPTIONS

1. Oil Resources

The geographical distribution of proved oil reserves as of the end of 1988 is shown in Table II-1. The minor position of the OECD group of countries, in terms of both reserves and the reserve-to-production ratio, is clearly illustrated. It is further aggravated by the declining production of the United States, United Kingdom and Australia. It is also noteworthy that world proved oil reserves two years earlier were almost 200 billion barrels less than the estimates given in Table II-1 with a world reserve-to-production ratio of 32.5. Political factors notwithstanding, this is a substantial increase in reserves, occurring mainly due to revisions by Iran, Iraq, Abu Dhabi and Venezuela.

Table II-1: **Proved Conventional Oil Reserves, end 1988[1]**

Area	Billion Barrels	Million Tons	Share of Total (%)	Ratio to Production (Years)
OECD	63.3	8 200	6.9	10.7
Africa	56.2	7 500	6.1	28.6
Asia and Pacific	19.5	2 400	2.1	19.5
Latin America	122.1	17 100	13.3	50.5
Middle East	571.6	77 300	62.4	100+
CPEs[3]	83.9	11 300	9.2	14.5
Total World	916.6	123 800	100.0	41.0

1. Reserves of shale oil and tar sands are not included.
2. "Proved" reserves of oil are generally taken to be those quantities which geological and engineering information indicate with reasonable certainty can be recovered in the future from known reservoirs under existing economic and operating conditions.
3. Centrally Planned Economies

Source: British Petroleum; 1989 (Ref. 30).

2. Natural Gas Resources

It is clear that natural gas is likely to play an important role if there is to be diversification of transport fuels in the timeframe of this study. Natural gas can be used as CNG or liquefied natural gas (LNG), or converted into methanol, gasoline or distillates. It is often seen as a more abundant source of supply than oil. Table II-2 illustrates this latter point. Thus OECD countries have 12.6 percent of proved gas reserves, compared to 6.9 percent for oil. In addition, the number of individual sources of supply is greater for gas than for oil and their distribution is different, giving consumers greater choice and hence reducing the possibility of control by cartels. The OECD reserve-to-production ratio for gas is almost double that for oil, despite the fact that exploration for gas has historically been much less than for oil; this is largely because natural gas is often found when the prime target is oil.

Table II-2: **Proved Natural Gas Reserves (end-1988)**

Area	Trillion cubic metres	Million tons oil equivalent[1]	Share of Total Reserves (%)	Reserve-to-Production Ratio (years)
OECD	14.3	12 900	12.6	19.0
Africa	7.1	6 400	6.5	100+
Asia-Pacific	6.2	5 600	5.5	63.1
Latin America	6.7	6 000	6.1	70.0
Middle East	33.4	30 100	29.9	100+
CPEs	44.2	39 800	39.4	51.7
Total World	111.9	100 800	100.0	58.0

1. Based on 1.11 trillion cubic metres = 1 000 million tons oil equivalent

Source: British Petroleum; 1989 (Ref. 30).

As exploration is costly, companies do not generally explore until they can perceive a future market; this tends to limit reserve-to-production ratios. Clearly, the price at which the gas can be sold is also an important determinant of exploration, and hence reserves. For example, considerable anticipated demand for methanol from gas could raise forecast gas prices and increase exploration in areas where prices are currently low. Clearly, an increase in gas price would also affect methanol prices and, in turn, demand for that fuel. Developing countries could play an important role in future gas markets, depending upon their levels of domestic demand for their own resources. These complex inter-relationships are addressed in an assessment of alternative fuels prepared by the Department of Energy (DOE) of the United States (Ref. 133).

Furthermore, when considering the impact of increasing natural gas demand, it is important to realise that international trade is much more limited for natural gas than for oil. Certainly there are pipelines in Europe and LNG shipment, especially to Japan, but most gas is both produced and consumed on a regional basis. Thus the impact of increasing demand needs principally to be looked at in terms of the reserves of consuming regions and individual countries within those regions. If methanol were produced from gas outside the regions where it was consumed, as seems likely, there would be no direct effect on gas prices in consuming regions unless there were already large scale LNG shipments to those regions (for example, Japan). However, gas prices in methanol-producing regions could move up due to increased demand on their gas, depending on the nature of contracts in place.

Indeed, methanol can be regarded as an "intermediate vector", enabling use of small and/or remote gas fields where pipelining or capital-intensive LNG plants and transportation of the gas are not economically justifiable. It is assumed throughout the present study that methanol or synthetic products would generally be produced in the country possessing the gas reserves. Shipping LNG for conversion would usually be more costly, and would not enhance energy security unless a nation particularly wished to have conversion plants on its own soil. On the other hand, domestic gas resources would probably be used for CNG vehicles, perhaps making use of small, low capital cost gas fields.

The question of the availability of low cost gas for methanol production has been discussed in recent studies by the American Gas Association (AGA) and the United States DOE. The main conclusions of these two studies are set out below.

2.1 Estimates by the American Gas Association

An analysis conducted for the AGA examined three cases for development of a fuel methanol market in the United States: a low case of 0.144 million barrels per day (2 billion gallons or 7.6 billion litres per year) in the year 2000; a middle case of 1.44 mbd; and a high case of 3.6 mbd (Ref. 39). These and the following data are expressed in terms of actual methanol volumes, not energy equivalent (1 barrel of gasoline is equal to 2 barrels of methanol on an energy equivalent basis). These cases are treated as increments added to base case demand. They are not intended to represent exactly any specific proposal, but rather to represent the order of magnitude of proposals presently under public discussion. For reference, United States average oil consumption was 16.5 mbd in 1987, of which 5.8 mbd were imported and 10.4 mbd were used in transport.

The AGA analysis concluded that an additional supply of 0.144 mbd methanol in the year 2000 (about ten world scale plants) could easily be met from non-OPEC or USSR associated natural gas that is currently being flared. It is assumed in this analysis that United States gas would generally be too costly to provide more than a small percentage of methanol fuel required.

The magnitude of flared natural gas on a global basis would initially appear to be sufficient to produce an additional 1.44 mbd of methanol, the middle case. However, it may not be feasible to utilise all of this resource for methanol production because of logistical and economic considerations. Moreover, production of this volume of methanol from flared natural gas would require dependence on OPEC and Eastern Bloc supply for about three-fourths of total supply.

Alternatively, the methanol could be made from conventionally produced natural gas from existing reserves. Current annual global natural gas production is about 2 percent of proven reserves, whereas oil production is 3-4 percent of proven reserves. If annual natural gas production were increased from 2 percent to 3 percent of proven reserves and used to produce methanol, the additional 1.44 mbd of methanol could also be obtained. However, such a uniform increase would again entail substantial dependency, with reliance on the USSR for 60 percent of methanol production and OPEC for an additional 25 percent. In practice, therefore, substantially higher depletion rates might be required in certain countries. The exact amount of global reserves that are low enough in cost to justify methanol plants is not known. Nevertheless, the availability of flared gas plus low cost natural gas from all sources could probably supply methanol to support a United States neat alcohol transportation fuel market of a magnitude approaching 1.44 mbd.

The high case of 3.6 mbd of methanol would require the use of new sources of natural gas. According to a number of authoritative estimates, conventionally recoverable global natural gas reserves not yet discovered are about three times as great as proven reserves. Commercial scale natural gas reserves are available in over 60 developing countries. A portion of these resources could potentially be allocated to the production of methanol to supply an international liquid transportation fuel market. However, it is necessary to take into account development of gas markets in many developing countries, as well as LNG projects. For example, there are indications that CNG could be an economic proposition in developing countries with domestic gas resources. Furthermore, the above discussion relates only to United States demand. If methanol demand increased in the United States, it might well also increase in other countries.

2.2 Estimates by the United States Department of Energy

Data are available in a paper by Difiglio which stems from part of the United States DOE alternative fuels assessment. This gives current methanol capacity in selected countries, excluding the Soviet Union, as 0.35 mbd (Ref. 48). When the current methanol supply is compared to the projected demand for the year 2000 of up to 4 mbd methanol, it is clear that fulfillment of the potential demand for United States fuel methanol implies a major expansion of the worldwide methanol industry.

Methanol demand of 4 mbd would require that 90 bcm (3.2 tcf) of annual world natural gas production be devoted to methanol. This represents 6.7 percent of total natural gas production in market economies. For further reference, the United States gas "bubble" was 125 bcm at its highest, and 90 bcm accounts for around 20 percent of current United States natural gas demand. In Western Europe, 90 bcm represents about 40 percent of current demand. However, except for CNG use, it is unlikely for cost reasons that the United States or Western European gas would be used to produce alternative fuels. This comment excludes Alaskan, Canadian Arctic and Norwegian Arctic gas, which do not have high opportunity costs as they are not connected to gas markets by pipelines.

Thus, it is more important to consider the impact that 90 bcm of additional demand would have on producing countries outside the United States and Western Europe. If they are not now producing LNG, the effect on their gas market might be small in the timescale of this study, assuming that the methanol plants were distributed among a number of different countries. The question of which countries might be seen as “reliable suppliers” is another issue, although having more diverse gas sources prima facie implies more security.

The 90 bcm figure relates only to United States demand for 4 mbd methanol. If the United States wanted more, or other countries took the same approach, demand would be correspondingly greater. In addition, demand for natural gas for domestic heating and for electric power generation may well increase substantially. For example, it has been forecast that the present level of 30 bcm for electric power generation in Western Europe could rise to 90 bcm by 2010.

Table II-3 shows Difiglio’s estimate of the long-run supply of methanol (Ref. 48). Capacity estimates are based on flared and re-injected natural gas which could be economically recovered. They do not include remote gas that is not yet being produced.

Table II-3: **Long Run Methanol Production from Flared and Reinjected Gas**

Country	**Capacity (barrels per day)**
Current Capacity	350 000
Mexico	108 000
Algeria	1 267 000
Canada	331 000
Arab Emirates	144 000
Bahrain	29 000
Saudi Arabia	432 000
Argentina	72 000
Brazil	43 000
Chile	72 000
Trinidad	58 000
Malaysia	43 000
United States	1 094 000
China	36 000
Burma	7 000
India	14 000
TOTAL	4 000 000

Source: Difiglio; 1987 (Ref. 48).

These estimates are made only for countries currently producing methanol. Several other countries or areas have significant gas reserves but are not currently methanol producers. Within the OECD, these include Australia, Norway (Arctic) and United States (Alaska). For each country with gas reserves, an analysis of the size of the reserve and the cost of making these reserves available as a methanol feedstock is being conducted as part of the DOE study.

2.3 The Studies in Context

The two main sources discussed above give rather different estimates of potential methanol supply from low cost natural gas. Excluding the USSR in both cases, the AGA analysis gives potential methanol supply as 1.4 mbd, while the United States DOE analysis indicates 4 mbd. To put the studies in context, independent estimates of methanol supply potential are presented in Tables II-4 and II-5.

The potential of selected countries to supply methanol from existing gas reserves is illustrated in Table II-4. The table's estimates are based on proven reserves, assuming that 2 percent of reserves are dedicated each year to methanol production. Excluding the Soviet Union as a potential supply source, estimated potential for methanol from gas reserves is 1.66 mbd. This is just more than enough, by itself, to meet the middle AGA demand case, but less than half of what would be needed to meet the DOE demand case. However, there is substantial additional potential for production of methanol from flared and reinjected gas, as already hinted above.

Table II-4:
Estimated Methanol Supply Potential from Natural Gas Reserves in Selected Countries

Country	Proven Reserves (billion cubic metres)	Potential Methanol Supply[1],[2] (thousand barrels per day)
Canada	2 637	119 000
Mexico	2 078	94 000
Venezuela	3 000	136 000
Argentina	771	35 000
Qatar	4 621	209 000
Abu Dhabi	5 180	234 000
Saudi Arabia	5 020	227 000
Kuwait	1 378	62 000
Algeria	3 230	146 000
Nigeria	2 476	112 000
Australia	2 300	104 000
Indonesia	2 464	111 000
Malaysia	1 472	67 000
SUBTOTAL	36 627	1 656 000
USSR	42 500	1 921 000
TOTAL	79 127	3 577 000

1. Estimates are based on 330 stream days per year and are given for each 2 percent of proven reserves; this is in addition to existing methanol production capacity.
2. This table refers to proven reserves; countries may have large remote undeveloped gas resources which might be used to provide much higher levels of supply than indicated. However, this cannot at present be quantified.

Source: Cedigaz (France); 1989 (Ref. 38).

As shown in Table II-5, the total amount of reinjected and flared gas is substantially greater than either the DOE or AGA study assumes to be recovered. The amount of gas flared has declined by roughly 80 bcm since the late 1970s as more uses have been found for this gas. Reinjection of gas, in contrast, has steadily increased, roughly doubling in the last decade. Unfortunately, as discussed above, only a fraction of the reinjected and flared gas may be available where methanol production is desired. Nevertheless, when these potential resources are added to the potential for methanol production from reserves, both the AGA and DOE assessments of supportable demand may be considered plausible.

Table II-5: **Trends for Gas Reinjected or Flared**
(billion cubic metres)

	1975	1980	1982	1984	1986	1988
Reinjected	78	113	149	166	176	209
Flared	173	164	111	100	100	92

Source: Cedigaz (France); 1989 (Ref. 38).

3. Other Resources

3.1 Very Heavy Oils

There seem to be no clear definitions of very heavy oils (VHOs), tar sands, and bitumens. Their characteristics are high density, viscosity (less than 10° API), low content of distillable components, low hydrogen to carbon ratio (less than 1.5), solid at ambient temperature, and usually composition high in inorganics, sulphur, and metals. For the purposes of this study, all will be referred to as VHOs.

VHOs cannot be recovered by drilling and pumping as with conventional crude oils. They have to be mined or treated in situ by heat to be separated from the rock material. Such VHOs are similar to vacuum residual oil from conventional crudes and must be further upgraded to pumpable liquids before becoming part-feedstocks to refineries or feedstocks to dedicated refineries.

Total world resources of VHOs are estimated to roughly equal those of conventional crude oils but are less well known and explored. The known reserves are concentrated in Canada and the Orinoco-belt (particularly Venezuela), as Table II-6 indicates, and so could be regarded as relatively reliable sources of supply. For comparison, Saudi Arabia's proven oil reserves are around 170 billion barrels.

Table II-6: **VHO Resources**
(billion barrels)

Country	Potential Ultimate Reserves	Total Resources
Canada	290	2500
United States	19	20-40
Orinoco	263	1 200

Sources: 1. IEA/OECD; 1986 (Ref. 5).
2. Government of Alberta; 1985 (Ref. 72).

Recovery has occurred since 1967 in Canada and has recently started in the United States. In 1987, Canada producedabout 300 000 barrels per day (bd) of crude oil from VHOs, of which about 200 000 bd were from mining operations using the Syncrude and Suncor plants, and the balance were from in-situ processing of deeper deposits. Current Orinoco production is about 100 000 bd. Forecasts by the Alberta Energy Resource Conservation Board indicate synthetic crude production rising to about 800 000 bd by 2005; the in-situ production in that year is forecast at about 150 000 bd.

3.2 Coal

Methanol can also be produced by established processes from coal, which is much more widely distributed geographically than oil or gas and is therefore less likely to be a constraint from an energy security point of view.

Only a few national governments (for example, Australia and the United States) collect and report figures on what may truly be called coal reserves. In most OECD countries, while producers of oil and natural gas are obliged to report the volume of their proven reserves to government authorities, coal producers are less often required to do so, and standards tend to be less rigorously applied. Hence, it is not possible to refer to a coherent set of data as in the case of oil and natural gas reserves. These limitations of coal endowment data have been recognised for many years. The efforts of the IEA to devise a new set of global estimates, based on a single systematic set of criteria rather than on disparate national estimates, are an important step in the direction of rectifying that situation.

Nonetheless, considering the mass of coal that is known to exist in seams that are technically mineable, physical availability is unlikely to be a constraint on development, at least within the next several decades. Whether the coal endowment will be able to support expansion of economic production at moderate growth rates (i.e., up to 5 percent per annum) is less certain. But if markets are open, past experience suggests the coal industry will continue finding and developing low-cost coal supplies for many years. Low grade coals can be used for methanol production, the most important factors being their location and mining mode and the availability of water for conversion.

3.3 Biomass

The biomass resource base for ethanol production substantially overlaps that for methanol production, including conventional wood, short rotation wood, herbaceous energy crops, agricultural residues, and municipal solid waste. But it also includes sugar and starch crops such as sugarcane, corn, and milo. The technology for converting starch and sugar crops to fuel ethanol is well developed and is already commercially established, although the industry is relatively small compared to the petroleum industry.

Biomass resources have the advantage of being renewable and may be abundant, depending on local circumstances, but they also have limitations and constraints as feedstocks for transport fuels. These relate particularly to the maximum quantities available and to collection problems, competition for other uses, seasonality of supply, and total energy requirements for their transformation into liquid fuels.

Assessments of the exact potential of biomass resources for motor fuel are hampered by the fact that statistical data are scarce for many countries. Most estimates must therefore be calculated from rather crude national statistics on forestry and agriculture. There are also great difficulties involved in trying to assess (without detailed local surveys) the accessibility of the different biomass resources. Such evaluation problems are most often encountered in developing countries. The abundance of available resources is, however, crucially dependent on local factors such as agricultural practices and crop productivities, and these vary considerably among the different countries of the world. It is important that world-wide resource assessments continue to be refined and that biomass accessibility be addressed in terms of the incremental cost of producing the next available unit of biomass.

Virtually all ethanol fuel produced today is derived from either a grain such as corn (the predominant feedstock in the United States) or from sugarcane (the predominant ethanol feedstock in Brazil). However, the ethanol production potential of biomass resources such as grains and sugarcane is limited by their possibly higher-value uses as food crops. Only special circumstances make these crops currently suitable for economical conversion to fuel ethanol. For ethanol fuel to make a significant contribution to world energy supply, wood and herbaceous plants would need to be utilised. These feedstocks are widely available, but the use of such feedstocks would clearly require proper management to avoid adding to existing problems of de-forestation.

4. Economics of the Options

4.1 Costs Using Technology Now Available

The economics of each alternative are usually the single most important determinant of when, or indeed whether, a particular fuel will enter the market. Thus the purpose of this section is to provide, on the basis of available literature, cost ranges for each alternative fuel. It must be emphasized that this analysis is based on existing technology. The fact that the costs of alternative fuels are higher than those of conventional fuels presents an opportunity for appropriately targeted R&D, not a reason for indiscriminate budget cuts. Accordingly, possible cost savings through improved technology are an important part of this study and are discussed in sub-section 4.2. In setting out these ranges, a number of underlying factors need to be made clear, as described below.

4.1.1 Production Costs

Production costs of alternative fuels are set forth in Figure II-1. Distribution costs are discussed afterwards. It is important to separate these two types of costs, for methanol, ethanol and CNG require substantial investments in new infrastructure and vehicles, while VHO products and gasoline or distillate synthesised from gas do not.

The assumptions relating to Figure II-1 are set out, to the extent possible, in Annex A. Some of the literature sources do not provide all the assumptions made. However, it is considered that Figure II-1 does provide ranges sufficiently representative of IEA Member countries' work in this field, and the ranges are themselves sufficiently broad, that plants built today would produce these fuels at costs falling within these ranges. All data have been converted to 1987 United States dollars on the basis of energy equivalent to a barrel of gasoline.

The conversion to alternative fuels requires large capital investments and may entail problems associated with applying unfamiliar technologies on a large scale in an uncertain market environment. Therefore, oil prices may have to rise significantly above these values, and moreover give the impression that they will remain there, to justify such a major commitment of resources. This could provide an advantage to plants of lower capital cost, such as relatively small-scale gas-to-CNG, gas-to-methanol, biomass-to-methanol or biomass-to-ethanol plants for regional use. There would also be competition from enhanced oil recovery of conventional oil resources. In addition, the relative economic efficiency of these technologies compared to promoting increased vehicle fuel efficiency would need to be examined.

In the case of CNG, production was assumed to take place by either of two basic methods. In the "slow-fill" method used by most private fleets, natural gas is compressed and fed directly to fuel storage cylinders in the vehicle. Compressors are sized specifically for their loads so as to minimise capital costs. The "fast-fill" system is better suited to applications where the vehicle cannot be taken out of service. Natural gas is compressed into a large-volume storage system at a pressure somewhat higher than the maximum permitted filling pressure of the vehicle cylinder. The vehicle is then connected to the storage system and refilled in a few minutes. The fast-fill system is somewhat more expensive, due to the cascade capacity and larger compressor required.

FIGURE II-1 : **Alternative Fuel Production Costs**

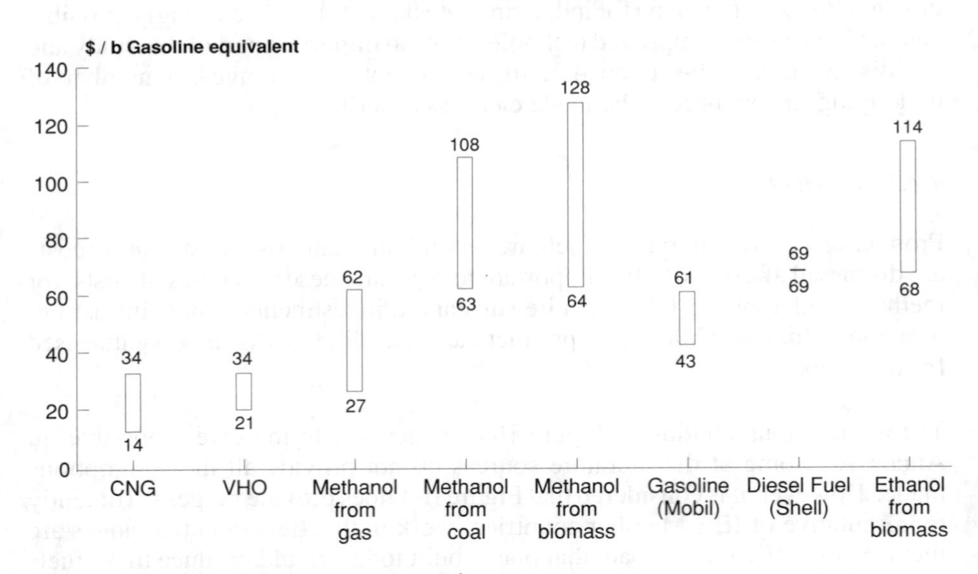

Sources: For details of data sources see annex A.
1. Gasoline equivalents calculated on basis of 2 : 1 for methanol and 1.5 : 1 for ethanol.
2. 1987 $US.

4.1.2 Distribution and End-Use Costs

It is important to address the question of the additional distribution and end-use costs associated with those fuels not compatible with the existing system, namely methanol, ethanol, and CNG. A major study by the United States Department of Energy (DOE) assesses additional distribution and end-use costs for alcohol fuels in some detail. For CNG, considerable analysis has been undertaken by New Zealand and Australia.

A paper by McNutt et al., which forms part of the United States DOE study, can be used to estimate the incremental costs of providing a methanol infrastructure for gasoline displacement in 2000 (Ref. 84). The calculation was done by simply dividing an estimated total of 2.83 billion barrels of gasoline displaced into the methanol infrastructure costs of $17 billion-$31 billion. This yields an incremental cost equivalent to $6-$11 per barrel of gasoline displaced. In McNutt's study, infrastructure costs include the additional cost of buying flexible fuel vehicles (FFV), estimated at $50 to $200 per vehicle. By comparison, the additional cost of pollution control equipment on present United States vehicles is estimated at around $400. It is not yet clear whether some part of the FFV additional cost could be traded off against reduced need for emission control equipment. Moreover, the infrastructure cost calculation does not take account of resulting benefits. McNutt's paper estimates that savings due to world oil price reductions resulting from methanol displacing oil demand would provide net benefits of between $7 billion and $10 billion in the year 2000, after taking account of infrastructure costs.

These calculations are relevant to the comparison between methanol and gasoline synthesised, via methanol, from natural gas. Synthetic gasoline would not require any additional distribution and use infrastructure beyond that already in place for gasoline derived from crude oil, representing a saving of $17 billion to $31 billion. However, the gross benefits should be the same, since in both cases the alternative fuels could be made from natural gas from the same sources and in both cases the same amount of conventional gasoline is displaced. Thus, in deriving overall costs for the two fuels, it seems reasonable to add $6-$11 per barrel to the cost of methanol but not to the cost of synthetic gasoline. Of course, synthetic gasoline will still have a distribution cost like conventional gasoline, but the present calculation is intended to try to represent the incremental costs of changing the distribution system, not actual values.

The costs of producing CNG, including the costs of both compression and storage and distribution, have already been included in the estimates in Figure II-1, and they assume that a gas pipeline system is already in place. In cities with reticulated gas supplies, CNG can easily be produced from natural gas supplied to re-fuelling outlets at existing service stations. It could also be produced in homes and businesses, again using the existing gas distribution network, if lower pressure (and hence lower cost) storage systems are developed. While pipeline supply could be costly to rural areas where natural gas is unavailable, the need for remote pipelines could be avoided through the use of dual-fuelled vehicles, trucking of gas from mother to daughter stations (as in New Zealand) or the use of small remote gas fields (perhaps in service mining operations). Taking all these considerations into account, it appears that any extension or increase of capacity for the distribution system would involve roughly the same additional production costs, regardless of whether it was for CNG or conventional gas. Thus, the incremental distribution and use cost of CNG is assumed in this study to consist essentially of the extra cost of vehicles required for its combustion.

A few original-equipment CNG-powered vehicles have been produced, and more are envisaged, but at present the vehicles are retrofitted. Based on New Zealand data for a passenger car vehicle conversion cost of $900 (1988 $US), a lifetime of 10 years and an annual CNG consumption equivalent to 2000 litres of gasoline, an incremental end-use CNG cost is calculated as $6 per barrel of gasoline displaced. As for methanol, this calculation relates to costs only. Australian and United States conversion costs are estimated to be higher, at least initially. They are in the range of $1000 - $1500, giving an incremental cost of $6-$12 per barrel. The Australian estimates of the capital cost of CNG conversion for a truck or bus is around $3500. However, the fuel consumption would also be greater than for passenger cars, so the incremental cost might be similar to that for cars. For CNG vehicles, DeLuchi has estimated that savings through reduced requirements for emission control equipment could be around $200-$400 (Ref. 47).

4.1.3 Overall Costs

The resulting overall costs for production, distribution and end-use are illustrated in Figure II-2. The calculations allow for the fact that CNG burns more efficiently in engines than do conventional fuels. Alcohol fuels have the potential to burn more efficiently, but have not yet reached sufficiently wide use to enable an equivalent level of precision in quantifying this benefit.

This addition of incremental distribution and end-use costs does not alter the overall economic ranking order, but does bring synthetic gasoline into closer competition with methanol. This competition might also involve non-economic factors which would favour synthetic gasoline, such as industry and consumer resistance to using a new product. These issues are discussed in Chapter V. Nevertheless, assuming both processes utilise the same low-cost source of gas, methanol might be about $13 per barrel cheaper than synthetic gasoline. Further, comparing the lower end of the ranges, VHO could be $10 per barrel cheaper than methanol. The end use implications of each of these fuels, such as the driving range limitations on CNG, are discussed in Chapter III.

FIGURE II-1 : **Alternative Fuel Overall Costs**

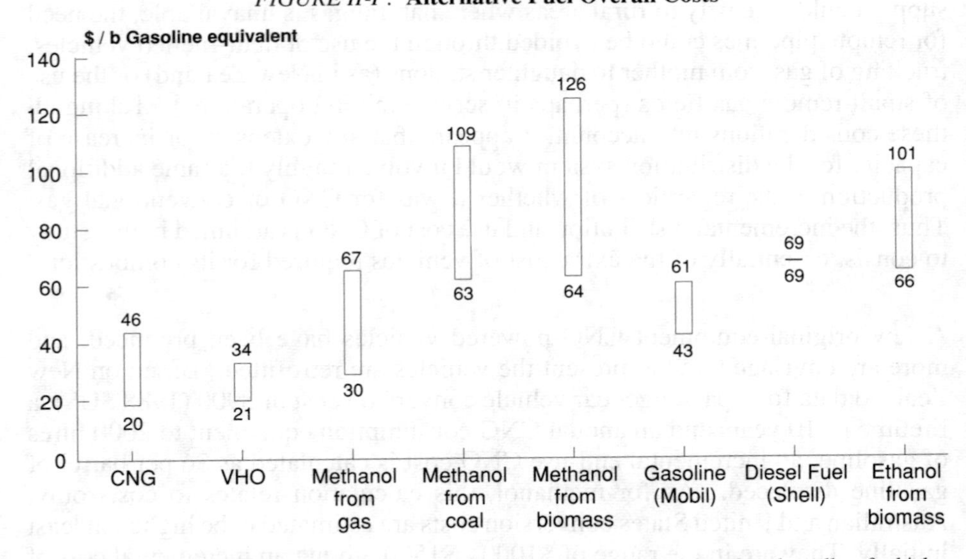

1. Gasoline equivalents calculated on basis of 1.8 : 1 for methanol and 1.3 : 1 for ethanol, to allow for higher efficiency of alcohol fuels in vehicules.
2. 1987 $US.

It should be noted that except for the lower end of the cost range for CNG and VHOs, all the alternative fuels are currently more costly than gasoline. Taking crude oil at $18 per barrel and applying a factor of 1.5 for typical refining and distribution costs (net of the value of other refinery products) gives a gasoline cost of around $27 per barrel. Only CNG and VHO may be competitive given today's oil and gas prices and today's fuel technologies.

The markets for oil and the alternatives are to a large extent coupled. As indicated earlier in this chapter, most gas reserves are located in countries which also produce oil. Therefore, it seems likely that those countries (or companies operating in those countries) would price any gas reserves being used as feedstocks for methanol production at a level which would maximise netback from both resources. For example, in a time when oil prices had risen to $40 per barrel, it seems unlikely that producers would price gas at a level which would allow a gas-derived methanol price of $27 per barrel, the minimum cost shown in Figure II-1. Rather, it seems likely that they would negotiate long term gas contracts linked to the price of oil, thus maintaining whatever competitive position alternative fuels may have. Those producing only gas could introduce a further element of competition.

In this situation, a further argument is that some alternative fuels, and methanol in particular, should be introduced for environmental reasons. Then a premium value would be imputed to the environmental benefits to the community of such fuels over gasoline. Again, this premium value appears difficult to quantify directly.

4.1.4 Impact on Refineries

Increasing production from VHOs is already having an effect on refinery operations. For example, it is estimated that current refineries in the United States will reach a saturation point of "bottoms" at a level of imports of VHO-derived crudes from Canada of around 500 000 barrels per day. In order to pipeline such crudes, they have to be about 22°API. This can be achieved either by upgrading in Canada or by addition of a diluent. It is certainly possible to refine heavy crudes using conventional technology, but only at a price. There is also a trade-off against refinery costs with decline in product quality, especially in the case of diesel fuel. But low-quality diesel can be used for heating oil and as residual fuel oil diluent.

To the extent that CNG or alcohol fuels replace gasoline or distillate, less crude oil refinery capacity would be required. As mentioned in Chapter V, few oil companies are at present interested in alternative fuels. Refineries would be involved if alcohol fuels were blended with hydrocarbons, to make, for example, M85 (which is 85 percent methanol).

In addition, an often important question involving distribution is the potential problem of ship-to-shore transfers of methanol and ethanol. These might use pipelines that, for safety purposes, are normally water-filled when not in use. Thus a change from distillate to methanol, for example, could involve a fairly major change in occupational safety and health practices and regulations relating to the handling of the fuel.

4.2. Potential for Cost Reduction through R&D

The market factors limiting R&D activity are well-known, and some continuing government role may be necessary until options have been clarified. At the same time, industry must be involved in market introduction, and it will perform more R&D when the market can be perceived. While the cost reduction issue seems a very important element to market introduction of substitute fuels, information on the actual levels of cost reduction achievable (and the R&D investments needed) is difficult to obtain. At the early stages of research, large cost reductions might be possible. Moderate cost reductions could also be obtained for a product or process as it enters the market and the learning curve is climbed. Finally, small cost reductions are available through detailed design improvements for well-established items.

4.2.1 Compressed Natural Gas

The production process for CNG consists of extraction and cleaning of the gas, then compression prior to end-use. Extraction and cleaning have been underway for many years and thus have little scope for further cost reduction. Although many compressors are already in operation, compressors of still higher efficiency, lower cost, and better reliability and durability, are desirable for fuelling stations. Lower-cost compressors are needed for home-use compression of natural gas to vehicle storage pressures.

4.2.2 Very Heavy Oils

Significant progress has been made in recent years in reducing the costs of production of fuel from VHOs in Canada. In 1982, decisions were made not to proceed with new projects on the grounds that they would not be economic, when crude oil prices were over $30 per barrel. Yet present estimates of products from VHOs range from $21-$34 per barrel for a new integrated mining plant, indicating aggregate gains through experience rather than any single major advance in technology.

Coal-oil co-processing is a relatively new process which has attracted considerable interest in the last decade. This process is an upgrading technology that involves the simultaneous conversion of coal and VHOs to synthetic crude oil. It therefore has the potential for creating new markets for inexpensive low rank sub-bituminous coals, especially where the coal and VHO resources are located close to one another. A study by the Canadian Energy Research Institute estimated the production cost of syncrude from co-processing at $29 per barrel (Ref. 86).

4.2.3 Methanol

Processes for producing methanol from natural gas have been improved considerably in recent years. Industry research is aimed at achieving a 10 percent cost reduction in the short term, with estimated 30 percent or even 50 percent reductions in the long-term. For example, a 1987 paper by the Davy McKee Corporation proposes a catalytic partial oxidation method to produce syngas, rather than the conventional steam reforming approach (Ref. 147). The catalytic partial oxidation method results in both lower capital cost and improved efficiency. Using the same assumptions as for Figure II-1, the production cost of methanol via the catalytic partial oxidation process is $30-$46 per barrel of gasoline equivalent, a reduction of 9-21 percent compared to the costs of a modern methanol-via-steam-reforming plant.

With regard to methanol from biomass, the United States DOE has a program to develop genetically-improved lignocellulosic energy crops that could be grown for $2 per million BTU (roughly $11 per barrel of oil equivalent), with a productivity of 8-9 tons per acre (roughly 20-22 tons per hectare). This could also be relevant to ethanol production. The gasification of biomass for production of fuel gas or synthesis gas as a boiler fuel has been researched extensively. Many gasifier designs have been researched at the bench and process development scales.

It appears that additional research could lead to incremental cost reductions for biomass-to-methanol systems. The greatest R&D need is the extended operation and testing of scaled-up systems to define for the private sector the technical and economic boundaries of these systems. Additional R&D is also needed on synthesis gas quality, gas clean-up technology and, to a lesser degree, methanol synthesis.

Biomass-to-methanol plants would be expected to have some characteristics different from coal to methanol plants. For example:

— lower financial risks and shorter construction time as the result of relatively small plant capacity;

— less environmental impact, especially as regards carbon dioxide;

— possible penetration of markets where other feedstocks are not available;
— ability to capitalise on short transportation distances of product to market;
— ability to use under-utilised farmland for fast-growing energy crops.

On the other hand, small plants lose advantages of economies of scale and more plants are required to make a significant impact on energy security. Whether farmland is under-utilised or not would necessarily depend on market conditions and could change over time. Finally, from an economic point of view, methanol from biomass is likely to be competing with methanol from coal towards the end of the time-frame used in the present study.

4.2.4 Ethanol

The technology for producing ethanol from sugar or starch based feedstocks using yeasts is well established. Plants of large scale for this technology (for example, approximately 4 000 barrels per day of ethanol) could be built in two years using readily available technology. R&D is continuing on improved feedstocks and production efficiencies, but it could be some years before results of this R&D are at a stage where they would be considered for commercial introduction. Moreover, further cost abatements are constrained by the maturity of the technology.

If ethanol from biomass is to become a major alternative at an attractive cost, it must broaden its resource base to include cellulosic as well as sugar and starch feedstocks. Accessing this abundant resource will require the development of new technology. Acid and enzymatic hydrolysis are the two technological routes that are currently being researched to provide a cost effective cellulose-to-ethanol technology. They should not only provide access to a greatly expanded resource, but should also result in a significant improvement in the economics of biomass-to-ethanol.

Cellulose-to-ethanol technology is now at the pilot-testing scale for some of the acid hydrolysis processes. For example, in New Zealand, work on scale-up was well advanced before oil prices fell and the programme was halted. In general, the research needs of acid hydrolysis processes are now primarily to scale up these systems and test them in continuous operation for both technical problems and production economics. The United States DOE is ending support for acid hydrolysis work because it believes that enzymatic hydrolysis will ultimately provide the most economic source of ethanol from biomass. The private sector continues to show interest in acid hydrolysis because it may prove to be an economic means of producing speciality chemicals from wood. Current costs from the two main acid hydrolysis methods are included in Figure II-1 and Annex A. Its cost edge notwithstanding, acid hydrolysis is also seen as a technology that is sufficiently close to commercialisation to be utilised for fuel production if circumstances warrant.

The more advanced enzymatic hydrolysis process known as simultaneous saccharification and fermentation is at the laboratory testing scale. Although this is a relatively immature technology, it shows potential for being more economical than acid hydrolysis. Enzymatic hydrolysis processes are potentially more efficient than acid hydrolysis processes, thus enabling significantly higher ethanol product yields and lower costs.

Simultaneous saccharification and fermentation has been under development for about ten years. Advances in the technology have been considerable, and the process shows significant potential for improvement. Currently, the Solar Energy Research Institute in the United States is studying the process at the laboratory and bench scale, and it has plans to produce a pilot-scale plant. Further research is needed to improve the economics of the process. Such research should focus on increasing the liquid fuel yield of all three major fractions of the biomass feedstock (i.e. cellulose, hemicellulose, and lignin). Biotechnological techniques are currently being applied to improving the micro-organisms and the enzymes they produce, which are critical to the conversion process. The Solar Energy Research Institute has estimated that the simultaneous saccharification and fermentation process, with today's technology, could produce gasoline equivalent ethanol for a selling price of $75 per barrel. However, a number of improvements to the simultaneous saccharification and fermentation process are anticipated which could lower ethanol production costs to around $40 per barrel gasoline equivalent. Ethanol production via bacteria is also being studied.

4.2.5 Synthetic Gasoline and Diesel Fuel

Methanol can be converted directly into gasoline using the Mobil Methanol-to-Gasoline process, or into gasoline and diesel fuel using a combination of the Mobil Methanol-to-Olefins and Olefins-to-Gasoline-and-Distillate processes. More specifically:

— the fixed-bed version of the Mobil Methanol-to-Gasoline process has been successfully demonstrated at the 14 500 barrels a day scale in New Zealand;

— a 100 barrels a day fluid-bed Methanol-to-Gasoline plant was tested successfully in the Federal Republic of Germany; this technology may have economic advantages over fixed-bed technology;

— the German pilot plant was modified to test the Methanol-to-Olefins stage for application in conjunction with the Mobil Olefins-to-Gasoline-and-Distillate process;

— the Mobil Olefins to Gasoline and Distillate process has been tested using a refinery olefinic feed in a wax hydrofinisher at a Mobil refinery in the United States.

The combined processes can produce up to 79 percent diesel fuel in the maximum automotive diesel oil mode and up to 84 percent gasoline in the maximum gasoline mode. Clearly, processes based on a first stage conversion to methanol will benefit from the improvements to methanol production processes discussed above.

In addition, Shell has developed its Middle Distillates Synthesis, a multistage process using natural gas as a feedstock. In the first stage, gas is partially oxidised to produce syngas, composed of hydrogen and carbon dioxide. In the second stage, called Heavy Paraffin Synthesis, the syngas is converted into heavy paraffins by the Fischer-Tropsch process. In the third stage, called Heavy Paraffin Conversion, the liquid products of the second stage are hydroconverted to middle distillates; the inputs and boiling range can be adjusted over a broad spectrum to yield the desired mix of products. Shell has decided to build a 12 000 barrel per day plant in Malaysia that uses this technology, to be completed by 1993 (Ref. 31). The plant will draw 100 million cubic feet per day of gas from an adjacent LNG facility. The $660 million invested in the plant is expected to be recovered through the superior quality of the distillates produced. These include up to 60 percent gasohol with no aromatics and a very high cetane number, or up to 50 percent kerosene with a very high smoke point. When such pure distillates are used, a smaller quantity is required in blends to yield a given amount of gasoline (Ref. 69).

Processes for conversion of natural gas to synthetic gasolines or distillates can be expected to experience significant cost reductions through R&D. Statoil of Norway has announced the development of a middle distillate process which is similar to Shell's, except that it works in a slurry reactor rather than in multi-tubed reactors surrounded by boiling water. British Petroleum has also developed a process to produce gasoline from natural gas; it has discussed siting of a 20 000 barrel per day plant in Malaysia, which might be built for operation in the late 1990s. Both BP and Statoil have indicated their processes would be economically viable at a crude oil price of less than $20 per barrel, as indeed the Shell process would seem to be (Refs. 31, 69).

The BP work, only parts of which have been made public, may exemplify the direct conversion of methane to hydrocarbons suitable for automotive fuels. In the shorter term gasoline production will still proceed through intermediate stages. Going more directly from methane to liquid fuels would substantially reduce costs, but this would be technically very difficult. Experts at a recent IEA workshop on chemical natural gas conversion (Ref. 19) concluded that direct conversion to higher hydrocarbons holds significant promise for future cost-cutting technical breakthroughs in the production of liquid fuels from natural gas. Catalytic research is expected to lead to improved methods for activating the methane molecule by breaking carbon-hydrogen bonds and inserting new atoms into the bonds. As progress is made in catalyst research, novel reaction pathways that result in the formation of higher hydrocarbons from methane may be identified. The most promising results from the university or government laboratories will be transferred to industry, which will concentrate on developing catalytic processes for the conversion of natural gas to desirable chemicals and fuels such as olefins, methanol, aromatics and gasoline.

As with all synthetic fuels, the optimum balance between gasoline/automotive and distillate/jet fuel will depend on specific national or regional needs. It should be noted that some R&D has the objective of converting methane to speciality chemicals of higher value than transport fuels. Consequently, plants constructed to make use of remote gas fields in some countries may in fact not produce fuels, at least initially.

Chapter III

VEHICLE END-USE CONSIDERATIONS

1. Vehicle Fuel Consumption Trends

The main driving forces behind the present assessment of alternative fuels are energy security and improved environmental protection. Energy security and environmental protection can be enhanced not only through increased use of substitute fuel, but also through reduced overall fuel consumption by vehicles. It is therefore necessary to make some comments on possible consumption trends.

As illustrated by Table III-1, which shows trends in the three major economies of the United States, Germany and Japan, technological improvements have led to considerable gains in fuel efficiency. The largest improvement has occurred in the United States, and this can be attributed largely to reductions in vehicle size.

Table III-1: **Fuel Efficiency of New Cars**
(Vintage Average[1])

Country	(litres per 100 kilometres) 1973	1979	1982	1985	1986	1987	(annual change) 1973-82	1982-86
Germany	10.1	9.6	8.3	7.5	7.3	7.6	-2.2%	-1.7%
United States	16.6	11.6	8.9	8.7	8.4	8.3	-6.7%	-1.4%
Japan	10.4	8.6	7.7	7.8	8.3	8.5	-3.3%	+2.0%

1. Including imported cars.

Sources: Verband der Automobilindustrie eV., Frankfurt, 1988.
IEA; 1984 (Ref. 12).
IEA; 1987 (Ref. 8).
IEA; IEA Member Country Questionnaires, 1988.

To take the example of Germany, in light of the total reduction of 28 percent between 1973 and 1986, it is clear that there has been substantial technological progress. However, the continuous fuel efficiency increases in the automobile sectors of Germany and Japan have been partially offset by a shift towards bigger cars and, equally importantly, by a change in driving habits towards more short-distance driving. This trend seems likely to be echoed elsewhere.

Moreover, there are also signs of slowing improvements on the technology side in recent years. Slowing efficiency improvements may be due in part to changing R&D priorities of vehicle manufacturers as emission-related problems have become more pressing than fuel economy. While prior to 1982 average fuel consumption of newly registered cars fell steadily in all three countries, this trend slowed down considerably in the United States after 1982, and in Germany and Japan it was reversed (Table III-1). In the United States, this development cannot be explained by trading-up effects, as the share of large cars in total car sales has been relatively unchanged since 1982. This trend points to a reduced rate of technical improvement, as measured by fuel consumption per unit of vehicle volume over a given distance. In part, this slowing is natural; with so much improvement already achieved, still further increases in fuel efficiency are increasingly difficult.

In light of the above points, the IEA Secretariat Mid-Term Energy Model projected that OECD vehicle gasoline consumption would grow at an average annual rate of 1.3 percent over the period 1985-2000, i.e. from 10.5 mbd to 12.8 mbd. This is for gasoline only; in certain countries, there could be substantially higher rates of growth in the use of diesel fuel. Australian government forecasts, which indicate a yearly rate of increase in fuel used for road transport of 1.7 percent through the year 2000, show a 6.3 percent annual increase for diesel fuel and a 0.5 percent annual decline for gasoline.

The question of whether to target all road transport for substitute fuels, or just trucks and buses, is touched upon at various points in this study. To put the question in context, in the United States some 22 percent of total road transport energy consumption is by medium and heavy trucks, with a further 1 percent for buses (Ref. 76). It would thus appear that programs which target trucks and buses only may be missing a large portion of the ultimate marketplace for fuel alternatives.

In summary, increases in energy security through continued technological improvements in vehicle efficiency may be slower in the future than the past. In addition, such improvements may be offset by changes in consumer behaviour. However, to the extent that improvements in end-use efficiency can be achieved, they will tend to complement programmes to develop substitute fuels.

2. Very Heavy Oils in Vehicles

Very heavy oils can be (and are) used to make products similar to those from conventional oil sources, although products from VHOs do have some special characteristics. For combustion of diesel fuel made from VHOs, only relatively minor changes to the vehicle fleet are required. Yet road diesel operation with conventional fuel injection systems on a 35 cetane number diesel fuel has serious problems. Fire ring wear has been shown to decrease engine life by 70 percent (from 10 000 hours to 3 000). Hence, a cetane number of 40 appears to be the lowest acceptable level for conventional Diesel engines.

Low cetane diesel fuels from VHOs have been evaluated in a number of engines by various researchers in Canada. Recent results indicate that wear problems can be reduced or eliminated with electronically controlled injection systems. Moreover, for locomotive Diesel engines, reductions in engine life are not as dramatic and are more than compensated by savings in fuel costs. Consequently, railways are using a lower cetane Diesel whenever possible.

The North American present minimum cetane number of 40 contrasts with a number of 50 generally used in some other countries and regions. Indeed, there is a trend towards higher cetane numbers for environmental reasons. This requires that diesel fuel from VHOs be upgraded or blended. One approach would be to reduce the upgrading requirements by selling two grades of VHO-derived fuel, one for transport and one for heating. This approach is likely to be used with conventional crude oils in the United States. It is based on an agreement between engine manufacturers and refiners for low sulphur fuel to help meet the 1994 particulate emissions standard. Another approach would be to blend VHOs with petroleum-derived high-cetane stock. In general, significant changes to fuel specifications require thorough proving before they will be acceptable to engine manufacturers and to consumers; a few adverse experiences could prevent adoption of a new specification.

R&D needs for VHOs may be summarised as:

— Continued research on combustion, particularly on how reduced cetane numbers (perhaps as low as 30) affect performance and durability, taking account of different engine configurations;

— Development of procedures for measuring ignition quality (presently done by cetane index), particularly for fuels derived from non-conventional crudes;

— Research into the relationship between chemical species and ignition delay;

— Continued development work on preparation and testing of ceramic components;

— Studies on the trade-offs between product upgrading methods and costs;

— Research on Diesel engines to increase tolerance to low-grade fuels and to reduce particulate and smoke emissions.

3. Natural Gas in Vehicles

Both CNG and LNG appear to have particular potential for countries with limited or declining oil reserves but substantial availability of low-cost natural gas, such as Australia and New Zealand. However, LNG is at an earlier stage of development. LNG is stored at a pressure of 8 to 9 pounds per square inch and a temperature of -162°C; the engine runs on boiled-off gas. Its main advantage relative to CNG is a smaller volume tank for a given energy storage, and this is offset by greater fuel system costs and complexity.

Natural gas can be stored on board vehicles as CNG in high strength cylinders at high pressure. A limiting factor is the relatively low energy content per unit volume of CNG, which will restrict the operating range of dedicated CNG vehicles unless their tank capacity is increased. The reduction in operating range with CNG can be substantial. Experience indicates that for CNG fuel tanks of the same volume as the conventional fuel tank, the converted vehicle will have one-third to one-sixth of the conventional range when operating on CNG. One way of overcoming this limitation is to use dual-fuelling. Table III-2 summarises the characteristics of CNG.

Table III-2: **CNG Compared to Conventional Fuels**

Advantages	**Disadvantages**
high octane number	low cetane number
broad flammability limits	low flame speed
gaseous state allows better distribution	gaseous state gives 10-15% power loss (recoverable through the increase in compression ratio)
reduced maintenance/longer engine life fewer deposits, no soot longer spark-plug life less lube oil degradation less corrosive fuel	reduced operating range due to added weight and greater space for fuel storage
explosive range narrower than for gasoline or LPG	more expensive fuel tanks

3.1 Otto Engine Vehicles

Full substitution of natural gas for automotive gasoline in a vehicle is possible by converting conventional petrol engines to the use of CNG. For example, Italy has about 250 000 CNG vehicles and 230 compatible service stations, while New Zealand has some 110 000 vehicles converted to CNG and 400 filling stations to serve them. In the United States, roughly 30 000 methane fuelled vehicles are in service, while Canada has 17 000 such vehicles operating in fleets. There is also considerable interest in non-IEA-Member countries, as illustrated by proposed CNG vehicle conversion programmes in Argentina (135 000 vehicles), Indonesia (100 000), Pakistan (21 000), and Thailand (50 000).

Otto engines are generally used in cars or light trucks, where the added weight of a CNG cylinder is proportionately quite large, approximately equal to the weight of a passenger. Due to range and re-fuelling limitations, almost all such vehicles are dual-fuelled, thus requiring two tanks and fuel systems. The vast majority are retrofit conversions. Despite these disadvantages, the experience in New Zealand, for example, is that reduced range has not been a major detraction from using CNG in light vehicles. Range is usually sacrificed rather than incur the expense of increasing storage capacity, although it should be noted that individual trip distances are relatively short in New Zealand.

Research is in progress to use adsorbent materials in the tank to reduce pressure to about 30 bar from the 200 bar used now, thus avoiding the need for high pressure compressors. In Canada, the storage of liquid hydrocarbons is also being studied. Many types of adsorbent materials have been considered, including activated carbons, zeolites, clays and phosphates. With activated carbons at pressures of 300 to 400 psi, the percentage of natural gas adsorbed can be 10 to 15 percent of the weight of the carbon. So far it has not been possible to find an adsorbent material which provides the same storage capacity of usable gas at the same cost, weight and volume as with high pressure tanks. Typical comparisons to date indicate that the total weight of storage with adsorbent materials may be double that of compressed gas at high pressure in steel cylinders. These will be about double the weight of compressed gas storage in aluminum cylinders wrapped with fibreglass, which in turn may be about 30 percent heavier than cylinders made entirely of plastic composites. Thus, the overall weight of adsorbent systems may be as much as five times that of storage systems built with the latest composites. Moreover, the volume required for the adsorbent system (with the materials evaluated to date) may be two to three times that required for the high pressure gas tank system.

CNG engine technology is less advanced than for gasoline engines. This may be due in part to its retrofit nature. For example, conversion kits which match the performance and reliability of the engine management systems on new cars are still not in widespread use. When they are, CNG engine performance will improve. Their introduction may involve original equipment manufacturers.

The energy efficiency of using CNG or LNG, compared to gasoline, is shown in Table III-3. Improved thermal and warm-up efficiency more than make up for the reduced economy resulting from added vehicle weight. However, these data refer only to total relative energy efficiency. The gaseous state of methane results in reduced volumetric efficiency, so a larger engine is needed if equal vehicle performance is required.

Table III-3:
Changes in Energy Equivalent Fuel Economy and Total Relative Energy Efficiency for Performance with Natural Gas Fuels which Equals Performance with Gasoline[1]

Gaseous Fuel	Fuel Tank Weight	Thermal Efficiency	Warm-Up Efficiency[2]	Relative Fuel Economy
CNG[3]	-6%	10%	3.3%	107.3%
CNG[4]	-6%	10%	3.3%	110.3%
LNG	-4%	10%	3.3%	109.3%

1. Changed displacement to equalise.
2. Based on the so-called "cold start cycle" defined by the United States Environmental Protection Agency.
3. Reduced operating range with standard steel tanks.
4. Reduced operating range with lightweight composite tanks.

Source: United States Department of Energy, unpublished data, 1988.

3.2 Diesel Engine Vehicles

Use of CNG in Diesel engine vehicles is at a less established stage than for Otto engines. A number of fleet demonstration trials are under way. Australia has around 50 buses and 150 trucks operating on CNG, including a trial beginning in Perth which will use aluminium/glass fibre tanks giving a 450 kilometer range. Other trials are in progress in Canada, Italy, New Zealand and Sweden. In New Zealand, the Aucklund Regional Authority decided on commercial grounds in 1988, after a 12 month trial, to progressively convert a large proportion of its Diesel bus fleet. There is also interest in Australia and Canada in the use of LNG in Diesel engine vehicles, particularly large trucks used in off-road operations such as mining. This work is at an early stage; one possibility is to make use of low-cost natural gas deposits that are near remote mine sites. From an energy security point of view, concentration of R&D on use of CNG in Diesel engines is indicated for countries where demand is forecast to grow quickly. There is also growing interest in North America in the ability of CNG engines to meet new particulate standards for buses in 1991 and trucks in 1994.

The operating characteristics of Diesel engines, especially their higher compression ratios, are such that they use the energy in natural gas more efficiently than do automotive gasoline engines. Indeed, they can obtain more power from natural gas than from diesel fuel, since compression-ignition engines normally run below the stoichometric mix. Some of this power increase can be attributed to mixed-fuel operation at mixtures richer than the diesel fuel smoke limit.

Since most Diesel engine vehicles are trucks or buses, the extra weight of cylinders has little effect on fuel economy. In addition, the trade-off between weight and gas storage capacity can move more towards capacity than with smaller vehicles. On the other hand, there will be offsetting costs arising from a reduction in payload capacity. For storage of 250 litres of diesel fuel or equivalent, a full tank weighs 250 kilograms, a full LNG tank weighs 290 kilograms, and the 11 conventional steel CNG cylinders which are required weigh 1150 kilograms when full. For buses, work is under way to design tanks into the roof structure to reduce the overall weight and the number of tanks.

Important issues relating to CNG use in Diesels are whether to use conversion kits or purpose-designed original equipment and, interactive with this issue, whether to use dual-fuelling or spark-ignition. The issue of conversion versus original equipment manufacture is largely one of production volume; truck engine manufacturers are beginning work on original-equipment Diesel engines which run on natural gas as they see a possible future market. For example, Australian trial fleets include original-equipment spark-ignition, converted dual-fuelled, and converted spark-ignition engines. Conversions have the advantage of penetrating the existing vehicle fleet, while engines purposely designed and manufactured for natural gas use are likely to be more efficient. A summary of the relative advantages of spark-ignition conversion and dual-fuelling is shown in Table III-4.

Table III-4:
Characteristics of Different Methods of Using CNG in Retrofitted Diesel Engines

	Spark-Ignition Gas Engine	Dual-Fuel Engine
Substitution for distillate	100% natural gas under all conditions	93% natural gas at fixe speed and load; 65% natural gas at varying speed and load
Conversion cost	high, usually new pistons and possibly engine head	moderate, basic engine the same
Electrical ignition system	necessary	not required
Reversion to full distillate operation	not possible	possible at full power or near full power
Combustion	simple pre-mixed deflagration	complex diffusion flame and pre-mixed deflagration
Engine control	conventional throttle	air-fuel ratio but may require throttle

Source: Milton Workshop on Diesel Substitution by Gas in Vehicles, Australia, 1987.

As substitution of natural gas for diesel fuel increases, it effectively increases the octane requirement of the Diesel engine. At high load or high levels of gas substitution for distillate, natural gas can cause substantial ignition delay knock in Diesel engines. At low loads, the problem is high levels of unburned hydrocarbon emissions. Because of these factors, as well as the critical control required over the load/speed range, mechanical fuel control is not likely to be satisfactory, and electronic control will be necessary. Furthermore, at high gas substitution rates, the reduced diesel fuel flow through the injectors may reduce life. This is due to reduced cooling of the injector tip.

Conversion of Diesel engines to spark-ignition gas engines may reduce the compression ratio and hence the combustion efficiency. But reduced efficiency is traded off against reduced pumping losses and enhanced fuel cost savings. At the same time, the quality of diesel fuel has been tending to deterioriate, providing some relative advantage to competition from CNG, as well as from alcohol fuels.

In conversion of Diesel engines to dual-fuelled operation, the engines are modified to accept a variable mixture of diesel fuel and natural gas. The two fuels must be delivered separately to the combustion chamber. A small amount of diesel fuel is always required for ignition, but operation is possible on either diesel fuel or a mixture with natural gas. R&D is proceeding into systems for controlling the mixture of diesel fuel and CNG under varying operating conditions. The conversion technology for mixed fuel operation is at various stages of adaptation and development for different four-stroke Diesel engines. Conversion technology for two stroke Diesel engines is at the developmental stage; for example, Australia is undertaking a major project for the development of a conversion kit for two-stroke Diesel engines in concert with gas suppliers and a major United States engine manufacturer.

R&D needs for natural gas use in road vehicles may be summarised as follows:

— Further development to reduce the volume and mass of natural gas storage systems, including development of lightweight tanks and tanks which fit more easily into the shapes of vehicles;

— Development of low pressure storage methods to enable use of home compressors;

— Development of electronic control systems for dual-fuelled engines, in order to better match the fuel-air ratio with the compression ratio and deal with ignition delay knock;

— Further development of engines which use natural gas fuel exclusively, through modifications such as supercharging, turbocharging, optimised ignition systems, and improved spark-control devices;

— Continued fundamental work on ignition, knock behaviour as a function of speed and air-fuel ratio, the effects of turbulence on flame speed, and the effects of variable natural gas composition on combustion properties;

— Development of lower-cost conversion kits, especially for Diesel engines, both naturally aspirated and turbocharged, which can take advantage of the on-board sensors and computers used with electronic ignition systems;

— Generation of operational and performance data for engines under standardised test conditions;

— Further R&D to reduce LNG fuel system complexity and costs.

4. Alcohol Fuels in Vehicles

The introduction of alcohols represents a major change in the correlation between fuel and vehicle. Alcohols are therefore discussed below in terms of how they can be used in purpose-designed vehicles, taking into account the potential for future development of these vehicles. The discussion begins with the general characteristics of alcohol fuels and proceeds to the specifics of methanol and ethanol use in Otto and Diesel engine vehicles.

Ethanol and methanol have similar combustion characteristics, so the alcohols which are selected for use in the various regions of the world will depend on resource availability and the economics of their uses as a fuel. However, as discussed in Chapter II, methanol can be made from a broader resource base than ethanol and is much less costly than ethanol when produced from natural gas. Moreover, methanol has the lowest molecular weight and the greatest oxygen content of all the oxygenated hydrocarbons and thus represents the greatest deviation, in both positive and negative senses, from gasoline. For these reasons, methanol is featured in the discussion.

Attributes of alcohols which affect engine performance and economy are summarised in Table III-5. Methanol has a high anti-knock index, or pump label octane number, of 99. Current United States unleaded fuels, by comparison, have anti-knock indices in the range of 87-93. Consequently, engines designed for methanol use have compression ratios which are two to three numbers higher than engines designed to use conventional gasoline. Methanol has broad flammability limits, high flame velocity and lower flame temperature than gasoline. This allows methanol engines to have more favourable lean combustion characteristics and lower NO_x emissions than gasoline engines. However, the potential reduction in NO_x emissions will not be fully realised if engine compression is increased to take advantage of methanol's high octane rating.

Methanol's high heat of vaporisation will improve volumetric efficiency and decrease compression work, which will result in increased power per litre of engine displacement. This in turn will mean a smaller engine for equal vehicle performance or improved mechanical efficiency through increased specific power. However, low vapour pressure will make cold starting more difficult unless additives to correct this are provided in the fuel and/or special provision is made to supply heat at appropriate places.

Table III-5: **Fuel Alcohols Compared to Conventional Fuels**

Advantages	Disadvantages
high octane number	low cetane number
broad flammability limits	low vapour pressure (also an advantage in reduced evaporative losses)
high flame speed	low energy per unit volume
high heat of vaporisation gives increased volumetric efficiency	high heat of vaporisation gives poor cold starting
no soot deposits in cylinders	increase in formaldehyde emissions
low flame radiation	low flame luminosity, more so for methanol than ethanol
water soluble (water can be used for fire extinguishing)	explosive properties of neat alcohol vapours in closed containers

Neat methanol (M100), unlike gasoline, can explode in a closed tank if enough primer energy is supplied. It also has a flame which is invisible in daylight and is toxic if ingested (although not much more toxic than gasoline, methanol is more palatable). It is possible that fuels which are not much different from each other in appearance but are of different specifications, such as the fuel for the Diesel type engine and the fuel for the Otto type engine, could be erroneously used. In order to avoid this, a simple method for identifying the different kinds of fuels by marker-tracer may be needed.

In summary, methanol has a number of favourable characteristics but also several limitations. This suggests that the optimal alcohol fuel of the future might consist primarily of methanol but contain 10-15 percent hydrocarbons as blending agents to improve cold start, wear, detection (a marker-tracer), flammability and flame luminosity. Neat methanol could theoretically have lower exhaust emissions than a methanol-hydrocarbon blend. But since M100 has lower energy per gallon than a blend, its use would require a larger fuel tank for the same vehicle range and some increase in fuel system weight.

4.1 Alcohol Fuels in Otto Engine Vehicles

The properties of alcohol fuels are favourable to improved efficiency of the Otto work cycle. In the Otto cycle, efficiency increases with increased compression of the air-fuel mixture before spark ignition, with increased air-to-fuel ratio, and with high flame velocity. Alcohol fuels can provide all these favorable characteristics. The extent to which they actually do so, however, may depend on the conditions under which modified or new engine concepts are introduced in the marketplace.

Otto engines designed to take full advantage of the combustion properties of alcohols are considerably different from those optimised for gasoline use. Retrofitting of gasoline vehicles was carried out in some early fleet tests, but retrofitting to obtain an optimised alcohol vehicle would be too costly to be justified. Essentially all development efforts are therefore directed toward design of engines and vehicles for new production cars. The development of alcohol-fuelled Otto engines and vehicles is being performed by many car companies: Volkswagen, Mercedes, and Porsche in Germany; Volkswagen in Brazil; Ford Motor, General Motors and Chrysler in the United States and Brazil; Mitsubishi, Nissan, Suzuki, Toyota and others in Japan; and Saab and Volvo in Sweden. Recent development work and prototypes seem to be directed to fuel-injected cars with three-way catalysts and to lean-burn/fast-burn engines with only oxidation catalysts.

Table III-6: **Comparison of Gasoline-Fuelled and Methanol-Fuelled Vehicles**

	Gasoline	**Methanol (M100)**
0-60 mph, seconds	17.5	14.4
City mpg, L/100km	24.0 (9.4)	12.1 (19.4)
Highway mpg, L/100km	39.0 (6.0)	20.4 (11.5)
Composite mpg, L/100km	29.0 (8.1)	14.8 (15.9)
Energy consumption, MJ/100km	258	253

Source: IEA/OECD; 1986 (Ref. 5).

Table III-6 shows performance characteristics for a typical automobile model, as a function of whether it is fuelled by gasoline or by methanol. The methanol vehicle gives better acceleration performance for the same energy consumption. However, as expected from the relative energy densities of gasoline and methanol, the methanol vehicle uses roughly double the final volume of fuel. The volume of M100 required in today's methanol vehicles is typically at least 1.8 times the volume of gasoline required in conventional vehicles to travel the same distance. Foreseeable improvements may allow methanol vehicles to use as little as 1.7 or 1.6 times the volume of fuel in conventional vehicles.

It should be noted that field tests often do not include collection of engineering data, which are usually obtained from laboratory tests. The latter examine a full range of power, cost, and emission conditions, from which a test fleet strategy is selected. Thus, until large-scale fleet tests have been completed, any conclusions about the comparative performance of methanol vehicles and conventional vehicles must be considered preliminary. Experience with alcohol fuel and CNG vehicles is minimal, compared to that with vehicles using conventional fuels.

In order to alleviate the problems involved in any transition from gasoline to alcohol fuels, considerable emphasis is now being placed on vehicles with engines which can automatically adjust themselves to run on any combination of methanol and gasoline from a single fuel tank without driver action. Thus, a vehicle previously filled with one fuel could be refuelled with the other, which from the consumer's point of view is preferable to the dedicated methanol vehicle until such time as methanol filling stations become widespread. There are various ways of describing such vehicles; the terminology "flexible fuel vehicle" (FFV) is used throughout the present study. This technology was developed in the United States by Ford in collaboration with TNO of the Netherlands. Other car manufacturers are also developing vehicles of this type. Such an approach has been made possible through the development of electronic fuel injection technology in the early 1980s. With this technology, an inexpensive sensor (optical for Ford Motor, capacitative for General Motors) determines the methanol-to-gasoline ratio being delivered to the engine and, together with other sensors, automatically adjusts the fuel injection system and ignition timing.

It is reported that the California Energy Commission will order about 5000 FFVs for delivery commencing in 1990 and is setting up a 50-site fuelling network. The two largest gasoline distributors in California, ARCO and Chevron, have agreed to make methanol available at some of their service station outlets to assist in the trial. The Commission hopes that FFVs will make up 30 percent of the car fleet in California by 2000. The United States Federal Government also may order 5000 such vehicles.

The disadvantage of such systems is that the engine can only be optimised for the selected compression ratio and cannot fully utilise the higher octane rating of methanol. If the engine is optimised for the higher octane fuel, there must be output restrictions when operating on the lower. Porsche experience indicates that although it is possible to design an engine which can burn both M100 and unleaded gasoline in a highly efficient and clean way, there are several problems yet to be resolved. Of particular concern is the durability of the fuel filter, pump and port fuel injectors when exposed to methanol over long periods. Cold starting at low ambient temperatures (below -15°C) also requires some further development for standard production, although progress has been made. The optical fuel sensor can reliably distinguish the alcohol content, but the accuracy decreases when approaching straight gasoline. Fortunately, the exhaust gas oxygen sensor provides the necessary compensation.

Reforming of methanol produces carbon monoxide and hydrogen; these products contain 20 percent more energy than does liquid methanol, and 14 percent more than vaporised methanol. This may be of interest for on-board reforming of methanol by the use of exhaust gas heat to recycle waste exhaust energy to release chemically bonded energy to the engine and so improve the efficiency. A methanol-decomposing system using exhaust gas heat is complex, and the practicality of vehicle on-board systems can be questioned. Thermodynamic analysis of reformed methanol systems has not yielded promising results.

Lubricants must be formulated specifically for use of methanol in Otto engines. Conventional lubricants for gasoline are not suitable. Oils adapted to alcohol fuels by modified additive packages have therefore been developed, and they are used in both commercial (Brazil and California) and demonstration alcohol car fleets. These oils are a major part of current methanol fuel fleet test programmes, have already shown good results, and will assist in solving engine corrosion problems. Anti-corrosion additives may also help to reduce engine wear. Although the development work to combat corrosion and wear is far from finished, it appears that engine durability over 160 000 kilometers of operation appears achievable with properly specified fuels and motor oil formulations for methanol fuel, at least barring severe winter conditions.

Thus, R&D needs for alcohol fuels in Otto engines may be summarised as:

— Demonstration of optimised dedicated vehicles, eventually using a “methanol engine”;

— Design, construction and demonstration of vehicles to take advantage of alcohol properties such as high knock resistance, broader flammability limits, higher flame speeds, lower flame temperature and emissivity;

— Design, construction, and demonstration of purpose-designed vehicles to overcome deficiencies experienced in adapted gasoline engines such as cold start and driveability problems, corrosion and non-metallic component deterioration, non-luminous flames, and explosive mixtures in the fuel tank;

— Formulation of lubricating oils that perform satisfactorily with both methanol fuels and gasoline, including mixtures thereof;

— Solutions to problems relating to poor lubricity, that is, pump and injector wear (unleaded gasoline has shown an ability to reduce wear);

— Further development of corrosion inhibitors;

— Identification of fuel components, such as hydrocarbons with low molecular weight or similar materials with a low boiling point, which could improve flame luminosity, reduce explosion risk, provide acceptable cold starting/driveability and add odour and taste to deter human contact and consumption (a blend with 15 percent unleaded gasoline appears able to perform these roles);

— Use of a coloured marker-tracer for mixture location in the fuel distribution network;

— Development of methods to detect water in fuel methanol;

— Development of means to measure and rate alcohol ignition quality.

Car manufacturers have indicated the following additional technical issues for FFVs:

— Fuel sensor reliability and durability;

— Fuel specification for winter;

— Long term durability of engine in cold climate;

— Formulation and durability of catalyst for aldehyde control.

4.2 Alcohol Fuels in Diesel Engine Vehicles

The proper functioning of a Diesel engine depends to a large degree on the auto-ignition property of the fuel employed, as measured by cetane number or ignition delay. If the ignition is delayed, for example by low temperature or low compression ratio, too much fuel will be injected before ignition. Engine pressure will then increase rapidly, producing "diesel knock". Unfortunately, alcohol fuels have poor auto-ignition characteristics.

Yet alcohol fuels remain an intriguing option for Diesel engines because they burn cleanly and have less of a tendency than diesel fuel to form nitrous oxides. Diesel fuel has a high flame temperature which favors NO_x formation. Also, diesel hydrocarbons have a high molecular weight, which hinders their complete combustion. Hence, their use favors the formation of soot, which can require power limitation and create emission problems. It appears that alcohol fuels would have fewer environmental problems than diesel fuel and yet could be used in a Diesel cycle with comparable efficiency.

There are several technical solutions to the use of alcohols in Diesel engines which overcome their poor auto-ignition properties. These solutions are of two basic types. One type involves modification of the fuel itself, with only minor engine adaptations. The other type involves more comprehensive engine and vehicle adaptation to the use of straight alcohols. Diesel engines specifically designed for alcohol use are not yet widely commercialised. However, field testing is under way in demonstration fleets of buses and/or trucks in the United States, Canada, Germany, Japan, New Zealand and Sweden.

There are two ways to modify alcohol fuels for use in Diesel engines. One is by the addition of ignition-improving agents. The other is through emulsification of the alcohol fuel in diesel oil. The latter course, however, does not enable complete substitution for diesel fuel. Hence, this study concentrates on the former.

Ignition-improving agents provide a means of allowing Diesel engines to operate on methanol and ethanol when the combustion air is not well pre-heated. Once a Diesel is started, it can operate on pure alcohols under high-speed, high-load conditions. But cold start, part-load, and transient operation will not be possible because of misfiring as engine temperature falls. This problem can be solved by the addition of more readily ignitable compounds. The use of organic nitrate esters in low-quality diesel oils is already a common practice to improve the cetane number. Quite high levels (5 to 10 percent or higher) of costly ignition improvers have to be used for reliable start, particularly in cold climates.

There are several ways to adapt Diesel engines and vehicles to allow the use of straight alcohols in spite of their poor self-ignition properties:

— Partial replacement of the diesel fuel by alcohol carburetion ("fumigation") of the air inducted to the engine (a two-fuel system);

— Duplication of the injection system to the cylinders, one for diesel-oil as pilot fuel to start the combustion and one for the alcohol as main fuel (a two-fuel system); and

— Ignition of injected alcohols by spark-plugs or glow plugs for complete replacement of the diesel oil (a monofuel system).

It can be argued that the spark-plug-assisted engine is no longer a "Diesel" engine. The ignition is not by compression, but it is similar to the Diesel cycle in that fuel injection continues after the initiation of the combustion by the spark. As in the Diesel engine, the energy release will, therefore, be controlled by the injection timing. The FFV concept is not relevant due to the lack of miscibility between methanol or ethanol and diesel fuel.

Materials problems have been experienced with fumigation in turbo-charged engines if the alcohols are introduced upstream of the turbocharger. Rapid erosion of the compressor blades by the liquid droplets occurs. This particular problem may be overcome by introducing alcohol downstream of the compressor or by totally vaporising the alcohol before it is introduced. But excessive piston and ring wear have also been observed, and little systematic work has yet been performed on the nature of these wear problems or how to solve them. More research is therefore required before alcohol fumigation can be judged a useful technique.

Injection of two fuels into the pre-chamber or directly into the cylinders was originally introduced to overcome problems with low-quality diesel fuels and allied emission problems. The concept has proven very suitable for alcohol usage in Diesel engines if fuel pumps and injectors are externally lubricated or re-designed to compensate for the poor lubricity of the alcohols. The system has the potential for very high replacement of the diesel oil, from 75 percent to 95 percent depending on the duty cycle. In fact, a 100 percent replacement would be possible if the high-cetane pilot fuel consisted of alcohols with a sufficient content of ignition improver to burn a number of fuels with low ignitibility as secondary or main energy fuel (for example, water-containing alcohols). The only problems reported from Germany and Sweden relate to injector needle tip seats, which eroded more rapidly because of the poor lubricity of the methanol. The drawback of the dual-injection engine concept is the increased cost of a second separate fuel system. Still, the cost over the lifetime of the vehicle may turn out to be considerably lower than for systems using modified fuels (ignition improvers) with little engine adaptation.

The most notable recent development of spark-ignited Diesel engines is the MAN direct-injection, stratified charge engine being field-tested within the alternative fuels programmes of Germany, New Zealand, and the United States (particularly California and Washington). Similar developments of this type of engine have been made by others such as Ford Motor, General Motors, and Komatsu. The Japan Automobile Research Institute has also studied spark-ignited, methanol-fuelled Diesel engines. The thermal efficiency can be as good as, or better than, the Diesel counterpart. The MAN engines have about 5 percent lower energy consumption for methanol over a test route. Apart from cost increases, market acceptability of spark-ignited alcohol engines is currently somewhat constrained by concern that increased service may be necessary because of the electrical system. Only the service records now being gained at the field demonstrations can give more certain information about this.

Alcohols, particularly methanol, may ignite on hot surfaces. This pre-ignition property is a limitation in Otto engines, but it can be turned into an advantage for direct-fuel-injection Diesel engines by the use of glow plugs for the normal ignition when contacted by fuel vapour. Glow plugs are already used to assist cold start in conventional Diesel engines. Development and test work have been reported from Brazil and the United States. A bus in Cologne, Germany ran well on a glow plug methanol engine for over a year. There is also interest in research on compression/ignition engines which operate on methanol at higher temperatures, utilising the heat-retention properties of ceramic engine components.

Field testing has started in various locations in North America with a 2-stroke Diesel engine adapted to methanol. Engine modifications include glow plugs to aid cold start and low-speed/low-load operation, as well as an electronic injection control unit. Vehicle modifications include installation of stainless steel fuel tanks and an electric fuel pump. With few exceptions, good driveability experiences are reported. However, the fuel consumption on an energy basis is still higher than that of the diesel oil counterpart except during idling. Cold start was not acceptable until a block heater was installed. There have also been problems with fuel pump wear, probably due to methanol's low lubricity, and with fuel injector contamination leading to power loss in the engine. Field tests are underway in Canada using 2-stroke and 4-stroke engines with glow plugs and improved piston design.

In summary, Diesel engines (especially turbocharged) require more development work than do Otto-cycle engines to utilise alcohol fuels. Several different technical routes are being taken. Each manufacturer is trying to optimise particular designs determined by its particular technology base. Early indications are that dual-fuelled designs and those using emulsifiers may be less favoured. Relevant to this is the fact that Diesel engines have quite long operating lives, so that alternative fuelling could involve both retrofit conversions and purpose-designed alcohol fuel engines.

Thus, R&D needs for alcohol fuels in Diesel engines may be summarised as:

- Development of higher-temperature compression/ignition engines to take advantage of methanol's high octane quality at high temperatures.
- Improved methods of general utilisation;
- Continued investigation into the use of ignition improvers in methanol;
- Research into various problems associated with methanol, such as lubricity, lubrication, fire safety and toxicity (as also recommended with respect to Otto engines);
- Additional development of durable spark and glow igniters;
- Determination of the effect of methanol-diesel blends on engine emissions, with emphasis on particulates;

— Continued engine durability testing and exhaust system design;

— Development of means to measure and rate alcohol ignition quality;

Eventually, both Otto and Diesel engines may be replaced by a new methanol engine designed specifically to operate only on methanol fuel to make the best use of its characteristics. This would occur when an assured, sizeable market had developed and methanol fuel was readily available to the general public. This engine could feature lean-burn, high compression, and active ignition. Though similar engines have been researched for petroleum-based fuels, an alcohol version is still only a concept. The use of FFVs for Otto engines could provide a transition to the realization of this concept.

Chapter IV

ENVIRONMENTAL CHARACTERISTICS OF SUBSTITUTE FUELS

Environmental factors provide a considerable part of the impetus for diversification of transport fuels, especially in relation to the introduction of oxygenated fuels. The extent of this environmental pull varies considerably among different IEA Member countries, but the broad trend towards stricter regulations seems likely to continue. Details of this trend, including current regulations, are set out in Annex B.

1. Very Heavy Oil Products

The production of VHOs has substantial environmental impact, particularly when using opencut mining operations. These developments have not involved upgrading to synthetic light crude because of a growing demand for heavy crude oil in the United States. Whether an expansion of VHO production would meet environmental opposition remains to be seen. Most of the resources are remote from large centres of population, which could be an advantage but might involve opening up access to wilderness areas.

Future Canadian and United States emissions standards will significantly restrict allowable levels of particulates — to less than the amount of lubricating oil combustion products in some engines. Hence, it may well be necessary to use anti-pollution devices even in engines burning a high-quality diesel fuel. Current investigations, which relate to diesel fuels derived from both petroleum and VHOs, examine how diesel fuel composition affects emissions and assess the use of trap oxidisers to contain and burn off particulates. Future research might productively focus on how various chemical species (especially the aromatics found in VHO-derived diesel fuel) affect both particulate emissions and NO_x levels. Combustion science indicates that improved combustion techniques could lead to significant soot reduction. Though not easily achieved, the development of means to ameliorate the environmental impacts of VHOs might well also apply to other new fuels and fuels from conventional crude oils.

2. Compressed Natural Gas

The production of natural gas is established worldwide and is not normally seen as an environmental problem, though there may be some adverse effects associated with leakages. As for all fuels, the exhaust emissions will be governed by many interacting factors determined by the engine design, the vehicle in which it is installed, and the driving pattern. No accurate prediction of emissions change can be made unless detailed vehicle specifications and driving patterns are obtained, but some general trends can be predicted.

Hydrocarbon exhaust and evaporative emission products will be the characteristics most changed by the use of CNG. Because methane is a much simpler hydrocarbon than those found in gasoline and mixes more uniformly with air, combustion will be more complete and theoretically will result in lower unburned hydrocarbon exhaust emissions. In practice, such emissions will depend on the particular engine in which methane is burned. Cold start emissions, which are a large part of the city cycle emissions from gasoline and methanol vehicles, should be reduced considerably compared to the liquid fuels at ambient temperature because a CNG engine does not have to be warmed up or choked.

Particularly with respect to evaporative emissions, it is important to note that methane is inert as far as atmospheric photo-chemistry is concerned. Thus, there is a question as to whether methane should be counted with reactive hydrocarbons in calculating the hydrocarbon level. However, unburned methane contributes to the greenhouse effect. Fortunately, in this regard, CNG vehicle storage produces no evaporative hydrocarbons. Fuelling and LNG vehicle storage yield negligible losses, except for limited LNG boil-off.

Overall hydrocarbon emissions from the use of natural gas fuel will depend to a large degree on the type of engine employed. Hydrocarbon emissions from CNG spark-ignition engines will be largely methane, but emissions from dual-fuel diesels will inevitably include diesel fuel. The hydrocarbon emissions from Diesel engines converted to the Otto cycle are producing higher total hydrocarbons than are their Diesel counterparts. Efforts to reduce these hydrocarbons are required.

Nitrogen oxides, the main photo-chemical precursors besides hydrocarbons, are the product of high temperature combustion with some excess oxygen present. Peak combustion temperatures in spark-ignition engines are generally lower for natural gas than for gasoline. But the natural gas mixture is often set to contain more excess oxygen than gasoline does. In this case, the use of CNG may cause overall NO_x emissions to increase. If ultra-lean burn engine fuelling is used to minimise excess oxygen, then the use of CNG should in principle cause overall NO_x emissions to be reduced. However, in the United States and Japan, where three-way catalysts are used to lower NO_x from gasoline cars, running "lean" or "ultra-lean" on CNG could increase NO_x because the NO_x reduction catalyst would not be effective. Moreover, if very high compression spark-ignition engines are used to take full advantage of CNG's high octane number, then NO_x may be significantly increased because of higher combustion temperatures. Catalysts for CNG engines have not yet been fully developed. Thus, depending on the balance of these trade-offs, nitrogen oxide emissions may be lower or higher for CNG spark-ignition vehicles than for gasoline vehicles.

In dual-fuel Diesel engines, the NO_x formation process is complex. It occurs in both the heterogeneous combustion zone, where CNG and diesel fuel are burned together, and the homogeneous combustion zone, where CNG is burned alone. With a combination of high CNG substitution ratios and high load, combustion may approach levels seen in a high compression spark-ignition engine, and NO_x emissions could be higher than in conventional Diesel engines. At lower loads, NO_x emissions from dual-fuel Diesel engines should approximate those from conventional Diesel operation.

Carbon monoxide (CO) is the consequence of combustion with insufficient air. Since CNG can be burned in much leaner mixtures (that is, with much less oxygen) than can petrol, the CO emissions in a spark-ignition engine should be small. Diesel engine CO emissions will be small whether from spark-ignition engines using CNG only or from dual-fuel engines using a combination of CNG and diesel fuel. However, overall CO emissions may be greater than with existing diesel fuel operation.

In general, when Diesel engines are converted to dual-fuel use without increasing power output, particulate emissions will be reduced. But if the control system for CNG and diesel fuel does not meter fuel correctly during accelerating transients, then some over-fuelling can occur and smoke will result. Both petrol and CNG spark-ignition engines produce negligible smoke. Table IV-1 summarises the above comments.

Table IV-2 gives an example of some observed emissions from CNG use in non-catalyst cars. Such data are particularly difficult to obtain as relatively little work has been done in this area, especially with respect to Diesel engines which are purpose-designed and manufactured for CNG use.

Table IV-1: **Effect on Emissions of Change to CNG Fuelling**

Engine	Non-methane Hydrocarbons	Nitrogen Oxides[1]	Carbon Monoxide	Smoke	Carbon Dioxide
Spark-Ignition: Compression ratio unchanged Ignition timing optimised	reduced to zero[2]	increased	reduced	none on either fuel	reduced
Spark-Ignition: Optimum compression ratio for natural gas vehicle Ignition optimised Ultra-lean part load	reduced to zero	large increase	reduced	none on either fuel	reduced
Dual Fuel Diesel	large reduction	complex effects - probably increased	no change	reduced	reduced

1. NO_x emissions changes can be altered by use of emission controls such as exhaust gas recirculation or ignition retard.
2. If pure methane is used.

Table IV-2: **Comparison of Gasoline and CNG Emissions**
(grams per kilometre)

	HC	CO	NO_x
Gasoline[1]	12.0	96.0	3.6
CNG[1]	1.6	4.8	1.2
Gasoline[2]	3.2	36.1	1.1
CNG[2]	2.1	2.4	0.7

Sources: New Zealand Liquid Fuels Trust Board; 1988
Environmental Directorate, OECD; 1986 (Ref. 2).

These data indicate that if methane is not included in the hydrocarbon value, a dedicated, optimised natural gas vehicle could meet United States CO and hydrocarbon standards without a catalytic converter or the associated equipment (Ref. 47). In the best case, it could also meet the NO_x standard with only exhaust gas re-circulation-type control, which costs in the order of $10 per vehicle. In some countries, this could result in up to a $440 reduction in the initial cost of the vehicle, relative to a gasoline vehicle, including any reduction in maintenance costs due to the elimination of emission controls. At worst, the natural gas vehicle would require a NO_x catalyst and, perhaps, some relatively inexpensive control devices or fuel and air adjustments to control hydrocarbon emissions. The initial cost in this case would be reduced by less than $200 relative to gasoline vehicle.

3. Alcohols

From the production point of view, the environmental impact of a gas-based methanol plant is comparable to that of an oil refinery with the exception of emissions of volatile organic compounds and related ozone problems. Siting of individual plants would therefore be subject to the normal environmental permitting process for the country in question. It seems unlikely that coal-based methanol synthesis plants would begin operating in the time-frame of the present study. Such plants would have a greater environmental impact than those based on gas, but again, conventional technology is involved.

Ethanol production from biomass is also established technology, although the widespread use of ethanol fuel would involve considerable scale-up. At present about 8 percent of United States gasoline demand is met by "gasohol" containing 10 percent ethanol. There can be environmental problems associated with high effluent levels from production facilities. This, combined with a need to achieve greater use of the feedstock material to improve plant economy, has led to additional steps in newer processes.

Notwithstanding the need for good environmental control of both methanol and ethanol production processes, most of the environmental interest in these fuels relates to end-use. Indeed, as noted at the beginning of this study, their perceived environmental benefits provide one of the main driving forces for their introduction. When their full production cycle is considered, biomass fuels may have virtually no net carbon dioxide output because the carbon dioxide produced in combustion can be stored into new biomass. Methanol can be used to fuel passenger cars in areas prone to ozone or carbon monoxide formation, as well as to power city buses whose large Diesel engines generate particulates and NO_x. In fact, the largest demonstration programme on the use of methanol fuel in passenger cars has been undertaken in California, where atmospheric ozone concentrations often exceed United States government standards.

4. Alcohols in Otto Engines

Since the chemical composition of alcohols is quite different from that of gasoline, the exhaust gases from alcohol combustion are also quite different. The unburned fuel consists mainly of alcohols and formaldehydes, with little or no benzene, other aromatics, polyaromatics, or olefins. Formaldehydes have high photochemical activity and may be mutagenic; they are thus of major concern in the design of alcohol vehicles. In blended "gasohol" fuels, non-methane hydrocarbon emissions are also of some concern, as they may form a larger fraction of the exhaust gases than they do of the original fuel. Particulate emissions, however, can now be reduced substantially through the use of catalysts.

Combustion of alcohols tends to produce fewer nitrogen oxides than combustion of gasoline. In large part, this is because alcohols have a greater internal cooling effect, owing to their high heat of evaporation, and a lower flame temperature. The possibility of leaner operation and spark retardation may further lower NO_x. This is so to some extent even at increased compression ratios, which generally tend to increase NO_x levels. Exhaust gas re-circulation and water addition, which are easy with alcohols because of their complete miscibility, are additional means to reduce NO_x generation. By proper selection of motor and operating parameters, it may even be possible to reduce NO_x to levels that would obviate the need for NO_x catalyst.

However, catalysts will remain necessary to cope with the other exhaust gases from alcohol fuel combustion. While unburned alcohols may be a small problem, the main concerns regard formaldehydes and possibly alkyl nitrites. The formaldehyde fraction is about 10 percent of unburned alcohol, compared to 1 percent of gasoline. An oxidation catalyst, which is less costly than a three-way catalyst, can reduce the formaldehyde level considerably. However, the effectiveness of oxygen catalysts in reducing formaldehydes in various driving modes is not known precisely. In particular, there is some concern that current catalysts may not fully remove formaldehydes during the cold start and warm-up phases of vehicle use. To help assess this problem, development of an on-line method for measuring the formaldehyde content of exhaust emissions has recently begun in Japan.

The United States Environmental Protection Agency's (EPA) proposed regulations for light duty methanol vehicles permit such vehicles to meet the same requirements as gasoline vehicles. In practice, the organic emissions from methanol may have roughly the same total carbon content as those from gasoline. However, their composition is such that methanol vehicles may produce as little as 10 to 20 percent as much ozone as gasoline vehicles. Very lean burn combustion should also result in decreased CO emissions in methanol vehicles without compromising reduced NO_x emissions. Methanol and formaldehyde will increase, but not to toxic levels. Formaldehydes can be reduced by use of catalysts to near gasoline levels, which is to say values well below those of present Diesel cars.

A 1987 paper by Moses of the United States DOE provides a review of methanol vehicles and air quality impacts, based on the data available from United States demonstration fleets (Ref. 94). Those vehicles were not set up to reduce emissions or achieve specific results, but only to ascertain whether they were within regulatory limits. Table IV-3 presents a sample of test results for current light-duty vehicles.

Table IV-3:
Selected Emission Rates of 1985 and 1986 Light-Duty (LD) Fleet Vehicles: US Federal Test Procedure [1]

Vehicle Type	Fuel	Emission Rates (grams per mile) HC	CO	NO_x	(grams per test) Evaporative HC
8-cylinder LD vehicle[2]	M85 Gasoline	0.13 0.20	1.53 0.83	0.46 0.55	1.65 -
4-cylinder LD vehicle	M85	0.26	3.10	1.00	
8-cylinder LD vehicle[2]	M85 Gasoline	0.23 0.19	1.10 0.37	0.64 0.45	0.71 0.72
4-cylinder LD Truck	M85 Gasoline	0.27 0.17	1.26 0.44	0.54 0.32	0.79 0.61

1. These data are believed to be representative, but in-service emissions data are so limited that this cannot be ascertained.
2. Vehicles from two separate fleets.

Sources: California Energy Commission et. al.; 1987 (Ref. 34).
SAE; 1987 (Ref. 112).

Moses noted that while the data are scarce, all values here are comfortably within U.S. federal government limits. Values for hydrocarbons reflect total loading; that is, a large proportion of the hydrocarbon figure for each vehicle type represents oxygenates, which are significantly less reactive (local ozone-producing) in the atmosphere than are the long-chain gasoline hydrocarbons they replace. Also, the levels reflect engine calibrations that are for the most part optimised for driveability or minimum emission levels. Of interest is the consistent finding that, while it appears possible to configure engines for reduced hydrocarbons and CO on M85, the reduction in NO_x is at the expense of higher CO. Conversely, it may not be possible to lower all three at once, since CO and hydrocarbons are oxidised best in excess air in the engine and in the catalytic converter, whereas NO_x is reduced most effectively by the catalyst in oxygen-poor conditions. However, this is true of any vehicle and is reflected in the use of engines operating near stochiometric mixture, where maximum engine-out emissions are obtained, rather than in the lean mode.

Design strategy is important in determining trade-offs, just as in gasoline vehicles. Further results, including information on formaldehydes, are provided in Table IV-4.

Table IV-4:
Methanol and Gasoline Vehicle Emissions: US Federal Test Procedure
(grams per mile)

Fuel and Vehicle[1]	UBF	CO	NO_x	Aldehyde
Methanol, TBI Vehicle A	0.16	2.29	0.72	0.044
	0.18	2.25	0.78	0.045
	0.34	2.68	0.84	0.055
Methanol, TBI Vehicle B	0.17	1.60	0.83	0.025
Methanol, MPFI Vehicle	0.18	2.25	0.52	0.028
Gasoline, TBI Vehicle	0.24	1.60	0.34	0.005
	0.19	2.25	0.33	0.003
Federal Standard	0.41	3.40	1.00	

1. TBI stands for "throttle body injection". MPFI stands for "multi-point fuel injection".

Source: General Motors Corporation; 1987 (Ref. 112)

The CO and NO_x emissions were generally higher with the methanol than with the gasoline vehicle. This result was in contrast to that expected based on steady-state tests. Most of the difference in three-way catalyst converter efficiency for unburned fuel, CO, and NO_x can be attributed to the slower converter light-off with methanol as compared to that with gasoline. This results from the lower exhaust temperatures experienced and fuel enrichment required for good warm-up driveability of the methanol vehicles as compared to those for gasoline. Formaldehydes were about ten times higher with methanol than with gasoline. When operating lean, formaldehyde levels increased, but unburned fuel levels decreased. However, no work on catalysts specifically for methanol had been done until recently, and no results have been reported to date. Thus, it is not yet possible to tell if methanol use will enable reductions in the cost of emission control equipment.

The potential importance of the fuel-flexible vehicle, or FFV, has been discussed in Chapter III. Emission results from Ford FFVs are set out in Tables IV-5 and IV-6. Some noteworthy features of these alcohol fuels emission data seem to be:

— For all vehicles (none of which were operating lean), the CO values are higher for methanol fuel than for gasoline, albeit still within EPA regulations in all cases;

Table IV-5: **1985 FFV Escort Emissions**[1]
(grams per mile)

Vehicle	Fuel	HC	CO	NO_x
TO50	Gasoline	0.25	1.90	0.63
	M85	0.19	2.10	0.52
EPA Standards[2]		0.31	2.40	0.74

1. The data given are the average of two tests on each fuel.
2. Standards for gasoline engines.

Source: Ford Motor Company, 1987 (Ref. 60).

Table IV-6: **FFV Crown Victoria Emissions**
(grams per mile)

Vehicles	Ratio	Axle Fuel	HC	CO	NO_x
T548	2.73	Gasoline	0.20	0.63	0.55
		M85	0.13	1.53	0.46
T500	2.73	E85	0.30	0.70	0.54
Canadian Fleet	3.08	M85	0.19	1.07	0.47

Source: Ford Motor Company; 1987 (Ref. 60).

— NO_x emissions are somewhat improved compared with gasoline vehicles under these conditions;

— Table IV-6 includes data for E85 (an 85 percent ethanol mix), which are harder to obtain than methanol fuel data. However, these data show worse hydrocarbon and CO and similar NO_x emissions compared to another vehicle of the same type running on gasoline.

Evaporative emissions (today regulated only in Australia, Japan and North America) have been tested for in the trials of alcohol fuels in the California test fleets. Carburetted cars failed to meet the 2-grams-per-test limit established under the Sealed Housing Evaporation Determination standard, but the limit was met by the cars with fuel injection. The capacity and lifetime of the carbon used in the cannister traps to the atmosphere seem uncertain and need to be investigated. Some Ford results on evaporative emissions for FFVs are set out in Tables IV-7 and IV-8.

Table IV-7: **Evaporative Emissions for 1985 FFV Escorts**
(grams per test)[1]

Fuel	Diurnal	Hot Soak	Total
M100	0.11	0.24	0.35
M85[2]	0.16	0.24	0.40
M85	0.29	0.41	0.70
M50	1.02	0.68	1.70
Indolene Clear	0.57	0.30	0.87
EPA Standards[3]	0.32	0.22	0.54

1. Emissions measured under the US Federal Test Procedures Sealed Housing Evaporation Determination Tests.

2. Blended with 15 percent Indolene Clear.

3. Standards for gasoline engines.

Source: Ford Motor Company; 1987 (Ref. 60).

Table IV-8: **Evaporative Emissions for 1986 FFV Crown Victoria**
(grams per test) [1]

Fuel	Diurnal	Hot Soak	Total
M85[2]	0.54	1.08	1.62
M100	0.79	0.81	1.60

1. Emissions measured under the US Federal Test Procedures Sealed Housing Evaporation Tests.

2. Blended with 15 percent Indolene Clear.

Source: Ford Motor Company; 1987 (Ref. 60).

The California Air Resources Board is to test FFVs in order to set emissions parameters for the vehicles and establish fuel methanol specifications, particularly in the light of possible future moves to use of fuel-grade methanol.

The United States EPA is in the process of setting standards to provide planning stability for the automakers. The proposed standards will in effect regulate a weighted sum of hydrocarbons, methanol, and formaldehyde. The numerical standards for the weighted sum of "organics" will be equivalent to the existing hydrocarbon standards for gasoline-fuelled vehicles in terms of the total amount of carbon which can be emitted in the form of partially burned or unburned fuel.

The CO and NO_x standards will be the same as for gasoline-fuelled vehicles. These standards will apply to methanol-fuelled vehicles and to FFVs when running on methanol. Data are not sufficient to indicate that either CO or NO_x emissions are changed with use of a dedicated vehicle or FFV. In fact, the EPA has found that at low temperatures, dedicated vehicles and FFVs with current technology may have poorer CO performance than do gasoline vehicles in the absence of low temperature standards for either vehicle type. Since this is currently an area of uncertainty, however, the EPA will allow States to assume equal CO and NO_x emissions for methanol and gasoline vehicles.

Thus, areas for further R&D in the use of alcohols in Otto engines include:

— Construction and operation of optimised vehicles and related collection of data from such test fleets, under different optimisation conditions, to enable emissions parameters to be set and/or trade-offs understood;

— Development of fuel methanol specifications;

— Development of an on-line method for measuring the formaldehyde content of exhaust emissons;

— Development of catalysts for use with alcohol fuels;

— Investigation of materials or improved products for use in absorbing evaporative emissions;

— Improvement of cold starting and warm-up emissions controls.

5. Alcohols in Diesel Engines

Diesel engine technology is of great interest in North America, where EPA particulate emissions standards are setting the pace for the development of clean engines. The main technologies competing to meet these standards are alcohol fuels and trap oxidisers for use with conventional (including VHO-derived) diesel fuel. Alcohol fuels may currently have the lead, as the Detroit Diesel Company has announced that it will sell only methanol engines to meet the 1991 United States bus standard. Although there are no particulate emission standards in Japan, there is interest in alcohol fuels as a means of reducing NO_x emissions from Diesel engines, since standards are not always reached in some large Japanese cities.

Discussion of the emissions performance of Diesel engines is complicated by the variety of approaches taken to alcohol fuel use in such engines (as described in Chapter III). For all approaches, the greatest difference relative to conventional diesel fuel use concerns the exhaust emissions. The unburned fuel emissions, which are mainly alcohols, ignition improvers, and formaldehydes, are very dependent on the ignition quality of the fuel and are, in fact, a good way to measure this quality. These emissions may be more critical at low loads.

Combustion of alcohol in Diesel engines has a number of superior qualities. Smoke or particulate emissions are very low with alcohol fuels. Theoretically, neat methanol is especially low in particulate emissions. Emissions of polyaromatics and monoaromatics are also very low. In general, NO_x emissions increase with increasing load but should be less pronounced than with diesel fuel, perhaps as much as 50 percent lower with optimised engines.

In a few respects, however, the use of alcohol in Diesel engines may be problematic. New emission compounds to consider are methylnitrites and ethylnitrites, which are mutagenic. The CO emissions, which are low from a diesel oil fuel engine, may be raised somewhat with the alcohol fuel at low loads if the ignition quality is not good enough.

Fumigation, the pre-mixing of fuel and inlet air prior to induction into the engine cylinders, has been investigated rather intensively in both single-cylinder laboratory engines and multicylinder engines. For well-designed and controlled fumigation systems, there seems to be general agreement that:

— NO_x exhaust emissions decrease with increasing amounts of alcohol (resulting from lower peak flame temperature);

— CO and unburned fuel exhaust emissions increase with increasing amounts of alcohol, albeit from a low level typical for Diesel engines;

— Particulate emissions decrease with increasing amounts of alcohol, but the biological activity of both the particulate and the soluble organic fraction increases; there is no clear explanation.

Few numerical data are available for alcohol fuel in Diesel engines. The information in Table IV-9 relates to buses taking part in demonstrations in California, which started in 1983. The MAN engine is a four-stroke unit with spark-ignition and an oxidation catalytic converter. The General Motors engine is a two-stroke which uses glow plugs as a starting aid and has no exhaust after-treatment.

Table IV-9: **Comparison of Diesel and Methanol Bus Engine Emissions over Various EPA Test Cycles**
(grams per brake horsepower-hour)

Pollutant	**Typical New Diesel Bus Engine**	**New MAN Methanol Bus Engine**	**New GM Methanol Bus Engine**
Particulates	0.57	0.04	0.17
Nitrogen oxides	6.25	6.60	2.20
Carbon monoxide	3.22	0.31	1.31
Total organics	1.61	0.68	1.28
Hydrocarbons	1.51	0.001	-
Methanol	-	0.68	1.13
Aldehydes	0.10	0.001	0.15

Source: United States General Accounting Office; 1986 (Ref. 139).

Both methanol engines produce very low particulate levels. Low particulate levels and the absence of sulphur permit the use of catalytic converters which, as shown by the MAN bus data, reduce total organics and carbon monoxide. However, the General Motors methanol bus produced high CO and unburned methanol emissions, which is an indication of incomplete combustion and the need for additional engine design work. Adding a catalytic convertor, which EPA believes should be mandatory, could also lower these emissions. It is important to note that engines running on diesel fuel cannot use catalysts, due to the high particulate levels. Engines running on methanol can use catalysts to reduce aldehyde levels. The General Motors methanol-fuelled bus also had low NO_x emissions. The next generation of General Motors engine is expected to have lower particulate and aldehyde levels, with fuel economy equivalent to its Diesel counterpart. No catalysts have yet been installed, though several are being evaluated. Such buses have been running in a test in New York.

In general, the emissions-related R&D areas for Otto engines also apply to Diesel engines. However, the greater diversity of technologies for alcohol fuels in Diesel engines complicates the situation somewhat. Each manufacturer will concentrate on its own technology. Current knowledge and experience do not seem sufficient to establish that any single alternative fuel option considered in this report has a clear environmental advantage. For this reason, more work needs to be done to clarify our understanding of environmental impacts.

6. Greenhouse Gas Emissions

Some work has been done recently on the relationship between global climate change and alternative fuels. DeLuchi et. al. evaluated emissions of carbon dioxide (CO_2), methane (CH_4), nitrous oxide (N_20), and other greenhouse gases from combustion of a variety of fuels, including the use of gasoline, diesel fuel, methanol and natural gas in highway vehicles (Ref. 46). Total fuel cycle emissions from the initial resource extraction to end use were estimated. The results relating to the fuels covered in this study are illustrated in Table IV-10.

Table IV-10:
Summary of Emissions of Greenhouse Gases from Alternative Vehicular Fuels

Fuel/Feedstock	CO_2-Equivalent Total Emissions[1] (billion tons per year)	Change in Emissions Relative to Petroleum (per mile driven)
CNG/LNG/Methanol/Ethanol from Biomass	0	100 %
Double Efficiency of Vehicle Fleet	0.668	50 %
CNG from Natural Gas	1.081	-19 %
LNG from Natural Gas	1.135	-15 %
Methanol from Natural Gas	1.293	-3 %
Gasoline and Diesel from Crude Oil	1.336	—
Methanol from Coal, 30% more efficient	2.026	+52 %
Methanol from Coal, Baseline	2.639	+98 %

1. All CH_4 emissions, as well as N_2O emissions from power plants, are converted to equivalent CO_2 mass emissions.

Source: Deluchi; 1987 (Ref. 46).

It is clear from the table that the use of coal to make any highway fuel would substantially accelerate CO_2 emissions, relative to the base-case use of petroleum. The use of natural gas as a feedstock would result in a very small to moderate reduction; reductions using CNG are greater than for methanol from natural gas. Significant reductions in emissions of greenhouse gases can be attained only by greatly increasing vehicle efficiency or by using biomass or non-fossil electricity as the fuel feedstock. In the case of biomass, the net impact will approach zero if CO_2-absorbing replacement crops are planted to sustain the methanol economy.

Chapter V

MARKETING ASPECTS

The study has identified the most promising alternative fuels and the areas where technology needs to be improved for cost or other reasons. These needs are not trivial, but they are not overwhelming. They can be solved with time and money, including some help from international collaboration. Bringing these fuels into the market on a large scale is another dimension of the problem.

The questions surrounding the introduction of new technology into the road transport market are complex, due to the large number of actors involved and the highly-optimised nature of the existing technologies, and could themselves be the subject of a complete study. Indeed, current studies by the United States Department of Energy are largely focused on market development rather than technology development. Furthermore, the approaches taken by individual IEA Member countries will inevitably differ significantly, even though they all espouse the basic principles of free market operation. Accordingly, the study has not attempted to provide an exhaustive treatment of the subject, nor to set out prescriptions for action, but rather has sought to identify the main issues involved.

Since this study stems from energy security considerations, including possible government policy measures, it is necessary to address the much-discussed question of the extent to which alternative fuels are capable of "capping" the price of oil and/or providing a deterrent to supply interruptions. One point of view is that building one or two alternative fuel (say methanol) plants and the necessary regional infrastructure would provide a "demonstration of intent" to oil exporting countries that oil and refined products cannot be marketed at a price above that of the methanol from those plants. Even so, the question remains of who would provide the investment for the methanol plant and pay for any incremental production costs above those of gasoline or diesel.

The other point of view is that to give credibility to this option it might be necessary to produce methanol on a much larger scale. A study by the United States DOE regards replacement of 2 million barrels per day of oil imports (forecast to total about 8 mbd in 1995) as being necessary to noticeably influence world oil prices. It appears that the price elasticity is such that each 1 mbd of oil replacement might reduce oil prices by about 1 United States dollar per barrel. To replace 2 mbd of oil would require roughly 4 mbd production of methanol, or about 10 times current world production. Almost 300 plants of the 14 000 bd size presently favoured would be required.

While it may be possible to build a considerable number of such plants in three or four years, availability of land, engineers and equipment would probably limit the degree of parallelism possible. It therefore might take 15 years to build these plants, during which time the oil market could fluctuate considerably. For reference, the nominal lifetime of an oil refinery is 20 years, and some 90 percent of all refining capacity has been constructed over the last 40 years. Furthermore, pressure on the chemical engineering industry would be exacerbated if other countries decided also to opt for methanol strategies, for example if oil prices seemed likely to reach, and remain at, appropriate levels. This could result in a production-derived constraint on the rate at which methanol could be brought into the market. It could also indicate that a transition to alternative fuels would be easier in a small market, as relatively few plants need to be built to make a significant percentage contribution to energy security. New Zealand is a case in point here.

Notwithstanding these issues, if a new fuel is to substitute directly for an existing one in a free market, the following criteria must be satisfied:

— it must sell competitively,
— it must perform comparably, and
— it must be readily available.

In addition, in order to penetrate the market, all of the stakeholders must be satisfied, namely:

— the fuel supplier,
— the vehicle supplier,
— the motorist, and
— government, with regard to public and national interests.

Although all of these groups must be satisfied to achieve market penetration, only a fraction of each category needs to participate, provided that fraction is sufficient to reach critical mass. Not all stakeholders will be affected uniformly or even favourably, and all may take different positions, see different opportunities, and be willing to accept different risks. Thus, for example, crude-rich petroleum refiners are not likely to be interested in alcohol fuels for replacement of conventional products; rather, they would wish to maintain the status quo, though perhaps accepting an eventual transition to a synthetic product essentially the same as present gasoline or diesel fuel. Companies at present marketing both oil and gas might also be expected to prefer CNG to the alcohol fuels. Chemical companies (for example present methanol manufacturers) have distinct marketing channels that differ greatly from the fuel distribution systems, and so they have limited interest, or face major obstacles, in promoting a new fuel.

Economic competitiveness, which represents the difference between the market value and the cost of production and distribution, is certainly the most important among the above conditions for market penetration. This aspect has been mentioned in Chapter II. From this starting point, the fuels discussed in this assessment can be divided into those which require changes to the present distribution and use system and those which do not. As far as distribution to the consumer is concerned, gasoline or distillates made from VHOs or natural gas are no different than those made from conventional oil resources, although in the case of VHOs there are substantial implications for refinery operations. The marketing issues with these fuels would be essentially related to economics, although there may also be environmental issues as discussed below. There are indications that conversion of gas to gasoline may be preferable in a small market, while methanol could be the preferred product for larger markets. Small markets tend to be associated with a low level of domestic automobile production; imported cars running on gasoline might be easier to obtain than those for methanol, at least initially.

This chapter concentrates on CNG and the alcohol fuels since they require changes to the distribution and use systems. This complicates their market introduction by involving more than economics. Additional questions include consumer attitudes and the lack of incentive to supply either fuel or equipment in the absence of the other.

More specifically, it appears that the following conditions must be met before large-scale introduction of these fuels can occur:

— *Price Competitiveness:* If the fuels are not already competitive with conventional fuels, or do not become competitive through technological improvements, they will not penetrate the marketplace unless taxation policies or other governmental policies are used to alter relative fuel prices. As for the question of energy security, individual governments must decide what level of oil imports is desirable and what policies might be needed to limit imports to that level. Clearly, energy security does not equate with self-sufficiency, and free-trade in energy supplies may well provide the most economic approach.

— *Fuel Production:* Natural gas (or other feedstock) supplies, either for direct use or conversion to methanol, must be reliably available, with production being increased in balance with demand.

— *Fuel Distribution:* Standardised quality fuel must be widely available. For a large country, this could initially be on a regional basis, in which case FFVs or dual-fuel vehicles would need to be available. For Western Europe, for instance, with its many national borders, it is important that agreement be reached on standard fuel specifications. As noted in the example above, it is not yet clear who would distribute the fuel. For commercial fleet use, which seems likely to be the main initial use for CNG, the distribution question is simpler. Safety and quality standards must match those of present fuels; a few adverse experiences could seriously hinder market penetration.

— *Vehicles:* Vehicles running on alternative fuels must attain at least the same high standards of driveability, reliability and durability and safety as existing vehicles, without being excessively costly. Indeed, they may have to have higher standards to overcome consumer resistance to new technology. Optimised vehicle design is highly interactive with the standard fuel chosen. It has been noted that alternative fuels tend at present to be regarded as "second best" compared with gasoline and diesel and that a principal objective of R&D on alternative fuels has to be to remove that concern (Ref. 49).

— *Emissions:* Although the theoretical advantages of the alternative fuels may promote their market penetration, it must be clearly established that these advantages can be realised in practice and that subsidiary problems such as formaldehyde emissions can be solved.

Analysis of the Issues

The United States DOE study is the most detailed available which addresses these conditions for market penetration. The DOE analysis does not assume any direct United States Government support to methanol production. Rather, it hypothesizes that circumstances or Government action may reduce the pump price of methanol below that of gasoline and create a market opportunity that oil or chemical companies will seize. Fuel production facilities would be located so as to provide the cheapest product (Chapter II). The study estimates that 10 percent of FFV drivers will use methanol when its price (on an energy equivalent basis) is equal to gasoline, 50 percent will use it at 6 cent per gallon advantage and 90 percent will use it at a 10 cent per gallon advantage. Apart from price factors, the availability of FFVs is also crucial; the DOE analysis indicates that service stations will offer methanol in response to the number of FFVs in the fleet, not in response to actual methanol utilisation (Ref. 133).

Similarly, once vehicle manufacturers were convinced of the government commitment to methanol fuel, FFVs would come into production. FFVs are essential at first since motorists will lack confidence in methanol fuel availability. Only when the methanol infrastructure was fully established would dedicated vehicles be expected to take over. The DOE study (which considers, but does not feature, dedicated vehicles) targets penetration by the year 2000 of about 90 million FFVs and over 300 000 dispensing units for M85 at service stations. Depending on the scenario chosen, FFVs would reach 100 percent of new car sales by 1997 or 2000.

The vehicle stock replacement question is also addressed by Dyne (Ref. 49). He notes that for automobiles in North America (which have a median lifetime of about 10 years), 30 percent market penetration could be obtained over 10 years if, from a standing start, 30 percent of new cars were to be methanol-fuelled. He sees this assumption as unrealistic and takes the view that 30 percent market penetration might take as long as 30 years.

In addition, the 1991 and 1994 United States EPA regulations for heavy diesel-powered vehicles may result in the use of methanol-powered buses and/or trucks, unless clean-up technology with conventional (including VHO-derived) fuels is more cost-effective. The FFV concept is not applicable to Diesel engines, as discussed earlier. However, as fleet vehicles are normally fuelled at a depot, dedicated vehicles are not the problem in fleets that they would be for the private motorist.

The United States Congress has passed a bill designed to encourage the use of methanol, ethanol and CNG in autos. The legislation provides automakers with incentives to make cars capable of burning alternative fuels. The enticements are offered through the Federal Corporate Average Fuel Economy (CAFE) standards. The largest credits are earmarked for autos capable of running exclusively on natural gas or on a blend of 85 percent ethanol or methanol with 15 percent gasoline. These cars would be given CAFE ratings based solely on the amount of gasoline burned. Take, for example, a car burning a blend of 85 percent methanol with 15 percent gasoline blend and getting 20 miles per gallon (mpg). This would be given a CAFE rating of 74 mpg based on fuel equivalency of 1.8 times as much distance per gallon of gasoline as per gallon of methanol (since $20/[.15 \times 1.8] = 74$). The fuel economy of FFVs is calculated by assuming that an FFV uses the alternative fuel one-half of the time. Its fuel economy is, therefore, a harmonic average of the normal fuel economy and the dedicated vehicle fuel economy. In the case of an FFV with a gasoline fuel economy of 20 mpg and a fuel equivalency of 1.8, the figure would be 31.5 mpg (since $1/[(1/74 + 1/20)/2] = 31.5$).

As noted earlier, a number of demonstration vehicle fleets are operating in various countries. These provide essential information on factors such as reliability, fuel economy and emission levels, which are then used to guide continuing R&D. Although such demonstrations are an essential prelude to market introduction, they do not themselves enable commercialisation on a large scale.

In terms of practical experience, recent developments in the New Zealand market for CNG in vehicles seem instructive. Statistics indicate that a fall-off in CNG sales started in the second quarter of 1986 and accelerated in the third quarter. By the latter part of 1987, sales had fallen by more than 20 percent from their peak. On inspection, it appears that the decline may be due to:

— Falling petrol prices, which reduced the margin between CNG and petrol prices, so that the number of vehicles worth converting to and running on CNG has declined (though despite the higher annual fuel consumption required to produce an economic return, the simple payback time will still be well under 2 years for a motorist driving 20 000 kilometers per year);

— The general climate of ready energy availability;

— Withdrawal of Government backing.

This has occurred despite the continuing known advantages of CNG as a vehicle fuel, which have been discussed earlier and can be summarisedas:

— Lower fuel costs with a potentially good rate of return on the investment;

— Greater security of supply for the motorist because the fuel is supplied to refuelling stations by gas pipelines rather than by rail/road tankers;

— Cleaner emissions;

— Reduced mechanical (including engine) wear and tear.

This may indicate that consumers are becoming more preoccupied with the less favourable aspects of CNG in vehicles:

— Reduced range (compared to petrol), hence more frequent refuelling;

— Fewer refuelling facilities;

— Extra space required on the vehicle for fuel storage (applies mainly to dual-fuelled vehicles);

— Up-front expenditure required for the conversion;

— Some power loss.

Alternatively, the downturn may be more attributable simply to consumers' wish to remain with familiar technologies. In any case, it illustrates the problems of introducing a new fuel into the marketplace.

To summarise, the large-scale introduction of fuels requiring changes to the distribution and vehicle infrastructure faces particular problems due to the complex nature of the industries involved and the number of consumers whose confidence must be won. Individual governments need to decide what (if any) action they see as necessary to overcome the barriers discussed above. Some of the barriers are purely economic. Others may involve market limitations where government action could be used to improve the transparency of fuel choices and facilitate the introduction of alternative fuels. At least four alternative approaches appear to be of potential interest:

— *No Action:* One approach is for governments to take no action, on the assumption that markets will operate efficiently enough for the private sector to make all the necessary investments when the time is ripe.

— *R&D Support:* A second approach is to support R&D which the private sector is not likely to do alone, thus helping to narrow the options so as to facilitate industry decisions on investment. It would not be necessary for all technological questions to be answered before market introduction began, although the points made earlier regarding consumer confidence would have to be satisfied. An initially negative response due to poor reliability, for example, could be very damaging to market penetration. The progress from demonstration fleets to large-scale market introduction could then rely on market forces.

— *Incentives:* A third approach, requiring a higher level of government intervention, is based on the assumptions that alternative fuel introduction is inevitable due to the location of world oil reserves and that the timescale involved is too long to rely on private sector action alone. In this case governments need to act soon to prepare the way for market penetration of alternative fuels, and they could consider promoting this penetration by introducing tax regimes or other incentives.

— *Environmental Valuation:* A fourth example would be based on environmental concerns, with an implicit assumption that the price of alternative fuels will always be higher than those of conventional fuels because of coupling effects. The alternative fuels would be assigned values in terms of their environmental benefits over conventional fuels. This is a difficult evaluation to make, and one likely to differ among countries. However, if a value relative to conventional fuels could be determined, industry could then focus its efforts. This may seem an unlikely approach, but in fact the world production of methyl tertiary butyl ether has risen from 0.3 million tons in 1980 to 4.2 million tons without any subsidy. Rather, the economic competitiveness of this product has been obtained by changing the market conditions through banning lead in gasoline.

There are, no doubt, other approaches, but the above four seem to be representative of the main current opinions. Governments considering action can learn from the practical experience of Brazil and New Zealand in promoting alternative fuels. Regarding timescales, some of the alternative fuels could come into increasing use in the next few years. For example, at present oil prices, CNG is economic provided that low cost gas is available. The main obstacles appear to be non-economic barriers, for example those which impede the introduction of CNG into truck or bus fleets even though they do not require large numbers of CNG outlets. Large scale methanol use seems unlikely to commence before the late 1990s. Apart from oil prices and environmental regulation, the main factor influencing the rate at which methanol penetrates the market will be the choices that governments make from among the four main policy approaches outlined above.

Chapter VI

CONCLUSIONS

The conclusions first briefly set out the main characteristics of each of the alternative fuels. These characteristics include energy security, economic, and environmental aspects, as well as the leading directions for future R&D and demonstration which seem logically to follow. On the basis of these characteristics, some more general recommendations are then drawn.

1. Characteristics of Substitute Fuels

1.1 Fuels from Very Heavy Oils

Main characteristics:

— Already in commercial use, on a moderate scale;

— Large reserves of feedstock are available in Canada;

— Overall cost is estimated at $21-34 per barrel of gasoline equivalent;

— No modifications to the distribution and end-use systems are required, provided that the fuel is upgraded to a sufficient degree;

— Depending on the degree of upgrading, these fuels may produce higher particulate emissions than do conventional fuels.

R&D directions which are indicated:

— Improved production and upgraded technology to provide better quality products at lower cost;

— Development of cost-effective trap oxidisers, more efficient than those of standard Diesel engines, in response to particulate emission regulations.

1.2 Compressed Natural Gas and Liquified Natural Gas

Main characteristics:

— CNG is already in commercial use on a modest scale; only a few vehicles are testing LNG;

— 14 percent of world gas reserves are in OECD countries; for economic reasons, initial CNG and LNG use would probably depend on use of indigenous resources;

— Overall cost is currently estimated at $20-46 per barrel of gasoline equivalent;

— For distribution, existing pipelines could often be used; for end-use, major vehicle modifications are required;

— Little data is available, but non-methane hydrocarbons, carbon monoxide, and particulate emissions are certainly lower than for petroleum-based fuels; questions remain about NO_x levels.

R&D directions which are indicated:

— Development of lighter weight and lower cost gas storage tanks; research into higher density storage phenomena;

— Cost reduction of engine conversion kits and optimisation of engine operation up to the standards of modern engines running on conventional fuels (this applies particularly to Diesel engines);

— Continuing development of engines designed from the beginning to run on natural gas;

— Investigation of whether cost savings are available through reduced need for emission control equipment, provided NO_x can be decreased.

1.3 Alcohol Fuels

Main characteristics:

— Methanol is at the fleet trial scale in several countries; ethanol fuel is in large scale use in Brazil;

— Much of the gas reserves in OECD countries would be too costly for methanol use. In general, gas is more widely distributed than oil, implying some additional energy security. Coal resources are widely distributed. Current indications are that for most IEA Member countries, ethanol from crops could not replace more than around 10 percent of current national transport fuel consumption. Alcohols from specially-grown forests and agricultural biomass might have greater potential, but this is unproven;

— Overall costs of producing gasoline-equivalent with present day technology are estimated (Annex A) at:

. methanol from gas,	$30-67 per barrel
. methanol from coal,	$63-109 per barrel
. methanol from biomass,	$64-126 per barrel
. ethanol from biomass,	$66-101 per barrel

— Major modifications to both the fuel distribution and end-use systems are required;

— Alcohol fuels should give lower levels of hydrocarbons, carbon monoxide and nitrogen oxides than do conventional fuels, but they produce higher levels of primary formaldehydes; secondary formaldehydes originating from volatile organic compound emissions decrease, with consequent reduction in ozone formation. From the limited data available, vehicles adapted to run on methanol do not provide clear net benefits at present apart from much lower particulate emissions from Diesel vehicles and much less reactive volatile organic compound emissions from Otto engine vehicles. However, these vehicles have not been optimised to give lower emissions.

R&D directions which are indicated:

— Technology improvement in the existing methanol production processes. Cost savings of 10 to 30 percent have been postulated. (Such improvements would also benefit coal- and biomass-based methanol production);

— New catalyst systems;

— New technology for syngas preparation;

— Direct synthesis from methane;

— Conversion of ligno-cellulose (biomass) feedstocks;

— For vehicles, manufacturers need to establish levels of reliability and longevity equivalent to those for conventionally-fuelled vehicles. This will involve continuing fleet trials; Diesel technology is considerably less advanced than it is for Otto engines;

— Development of standard fuel formulations having the maximum alcohol content permissible from operability and safety standpoints. This is needed both for cross-border traffic and to facilitate engine design;

— Development of additives that improve flame luminosity;

— On emissions, further engine-optimisation and catalyst development should determine whether emission control costs can be reduced below those for conventional fuels;

— In the longer term, research into the purpose-designed "methanol engine" could lead to its replacing both Otto and Diesel engines.

1.4 Synthetic Fuels from Natural Gas

Main characteristics:

— Synthetic gasoline (Mobil process) is technically successful at a commercial scale in New Zealand; other processes are at laboratory to pilot plant stages;

— Gas reserves are the same as for alcohol fuels;

— Overall costs of synthetic gasoline and diesel are currently estimated from published sources (Annex A) at:

. gasoline (Mobil process), $43-61 per barrel

. diesel fuel (Shell process), $69 per barrel

— No changes to the distribution or end-use systems are required;

— Emissions essentially as for petroleum-based fuels.

R&D directions which are indicated:

— Methods of converting methane more directly into gasoline or diesel fuel offer the promise of major cost reductions. Because no infra-structural changes are required, such cost reductions could have a particularly significant effect on the market for transportation fuels.

2. General Recommendations

Despite current oil market conditions, there is once more increasing interest in diversification of transport fuels away from the sources used at present. This interest is driven by energy security considerations, particularly in the light of declining levels of oil self-sufficiency in a number of IEA Member countries. Environmental factors provide an additional impetus in some countries for the use of alternative transport fuels. However, as illustrated by the previous conclusions, fuels differ in their energy security and environmental benefits. Thus, any government support needs to be clear in its aims. For example, VHO resources are located in less politically sensitive areas than is much of the low-cost gas which might be used to make methanol. Further, the products of VHOs are cheaper than methanol and do not require infrastructure changes. Conversely, methanol promises environmental benefits over VHO and conventional petroleum products.

Natural gas is involved in most of the alternative fuels considered here. Whether there is sufficient low-cost gas to support conversion to transport fuels at a scale large enough to significantly enhance energy security is not clear. The importance of this area appears to justify further study.

The production of alcohols from biomass is in many ways a land-use question, with many implications for agricultural and forestry policy. The study does not address the costs and benefits of farm subsidies, but any government considering support for large-scale production of alcohols from biomass should note that the real cost of the product is likely to be around $60 per barrel of gasoline equivalent, or more.

With the exception of the lower end of the CNG and VHO product ranges, the current estimated ranges of overall costs for the alternative fuels are all above those of conventional fuels. The greatest potential for cost reduction is probably through improved production processes on a larger scale. On the basis of the limited information available, it appears that potential cost reductions would be greatest for synthetic fuels from gas, then alcohols from ligno-cellulose biomass, followed by methanol from gas, then VHO products. CNG does not have a production process in the same sense, although there may be some scope for limited cost reductions through better compressor design. This cost reduction potential should be seen in the light of the cost estimates using existing technology.

End-use cost reductions may not have as large an effect as will process improvements on overall costs per barrel, but capital costs are important to the vehicle purchaser. Thus reductions of perhaps a few hundred dollars in the cost of a CNG conversion or in the additional cost of a methanol flexible fuel vehicle are worthwhile targets. It is also valuable to clarify the possibilities for trade-offs between the additional cost of such vehicles and reduction in emission control equipment. In this context, it should be noted that it is likely that emission regulations will become increasingly stringent, thus effectively raising the cost of conventional fuels.

Although there remain many areas where technological development is needed, diversification of transport fuels is to a large extent a market development question. It is essential, in introducing alternative fuels, that economic resources be allocated to activities where they can contribute most to national output. This will remain true notwithstanding any perceived urgency to redress declining self-sufficiency in crude oil. Thus, only individual countries can decide what level of oil imports and/or forecasted oil prices may justify government promotion of the introduction of alternative fuels. Particular marketing problems include:

— Fuel price, availability, standardisation and safety;

— Vehicle price, performance, optimisation, safety and emissions.

Economic viability and environmental acceptability are not the only criteria for successful market penetration. It will also be important that these alternative fuels not be perceived by the general public as second-best to existing fuels. In addition, it may be that promotion of increased vehicle efficiency would be more cost effective than promotion of alternative fuels. However, the latter, while appearing to be an area deserving further study, may prove to be a complement to, rather than a substitute for, a policy to limit and diversify transport sector fuel consumption.

The timing of the introduction of the various alternatives, seen against the background of the factors discussed above, might be as follows:

— CNG is already in use to a limited extent. For countries with low-cost local gas resources, this use could increase substantially in the early 1990s. However, previous experience shows that there are undefined market barriers above penetration of a few percent. These barriers need further investigation;

— VHO-based fuels are also in limited use; wider use may become cost-effective in the early 1990s. As in the case of conventionally derived diesel fuels, there could be problems with emissions unless cost-effective particulate traps can be developed;

— Methanol fuel, made from natural gas, may begin to enter the market on a significant scale in the late-1990s, but it has to overcome cost barriers and confirm its promise of environmental benefits;

— Ethanol fuel seems unlikely to be able to compete economically with methanol from natural gas in the time-frame of this study, except perhaps in specific local circumstances where, for political and social reasons, it benefits from economic subsidies;

— Synthetic gasoline and diesel fuel are more costly than methanol, but because infrastructure changes would not be necessary, major process cost reductions could lead to relatively easy market penetration. For this reason, government involvement may not be necessary. Individual country circumstances could be important and market share could increase around the end of the century.

Thus, any substitution by alternative fuels of a large percentage of present demand would have to be spread over two or more decades, and the investment required would have to be made almost entirely by industry. However, smaller percentage substitutions might still be sufficient to influence oil prices. If substitution by the alternative fuels in this study is slower than that indicated above, perhaps due to continuing discoveries of low cost oil, it is possible that there would be no "window of opportunity" for such fuels. Rather, the major problems yet to be overcome for hydrogen-powered or electric vehicles may have been solved, resulting in a transition directly to these vehicles from petroleum-based fuels. This is another possible area for further study, especially given concerns over global climate change.

In light of these findings, there appears to be no immediate justification to recommend a programme of enhanced production of alternative fuels. Rather, a watching brief should be maintained across the whole spectrum of the options considered in this report. In the meantime, it may be noted that international collaboration through the IEA is already active on the use and production of alcohol fuels and biomass conversion. The areas for R&D identified above are generally of interest to several countries. Thus, the existing collaboration should continue and be extended where appropriate. Areas for collaboration on other alternative fuels could include:

— CNG and LNG storage in vehicles and use;

— VHO upgrading and use;

— Advanced conversion of natural gas to liquid hydracarbons;

— Exchange of information and experience on test fleets using alternative fuels.

REFERENCES

A. OECD AND IEA REFERENCES

1. OECD; *Cities and Transport,* 1988

2. OECD; *Environmental Effects of Automotive Transport,* 1986.

3. OECD; *Transport and the Environment,* 1988

4. IEA/OECD; *Energy Technologies for Reducing Emissions of Greenhouse Gases* (Vols. 1, 2), 1988.

5. IEA/OECD; *Alcohols and Alcohol Blends as Motor Fuels, State of the Art Report* (Vols. 1, 2, 3), 1986.

6. IEA; *Coal Information 1988,* 1988.

7. IEA; *Energy Balances of OECD Countries 1986/87,* 1989.

8. IEA; *Energy Conservation in IEA Countries,* 1987.

9. IEA; *Energy Demand Analysis Symposium Proceedings,* October 1987.

10. IEA; *Energy Statistics 1986/87,* 1989.

11. IEA; Executive Committee of Implementing Agreement on Alcohols and Alcohol Blends as Motor Fuels, *Level II Report,* 1987.

12. IEA; *Fuel Efficiency of Passenger Cars,* 1984.

13. IEA; *Natural Gas Prospects,* 1986.

14. IEA; Standing Group on the Oil Market, *Petroleum Exploration in Developing Countries,* 1988.

15. IEA; *Production of Alcohols and Other Oxygenates from Fossil Fuels* (IEA Implementing Agreement on Alcohols and Alcohol Blends as Motor Fuels, Annex IV), 1988.

16. IEA; Committee on Energy Research and Development, *Report of Expert Group on Diversification of Transport Fuels,* 1987.

17. IEA; *Synthetic Liquid Fuels,* 1985.

18. IEA; Standing Group on the Oil Market, *World Methanol Industry: Use in Transportation Fuels,* 1988.

19. IEA; *Workshop on Enhanced International Collaboration on Chemical Natural Gas Conversion,* Oslo, Norway, 17th-18th November, 1988.

B. OTHER REFERENCES

20. *Alcohol Week;* various issues, 1987 and 1988.

21. Australian Department of Resources and Energy; *Review of the Production of Ethanol Sub-Programme,* 1985.

22. Australian Minerals and Energy Council; *Report of the Working Group on Alternative Fuels,* 1987.

23. Ban S.D. (Gas Research Institute); "An Agenda for Cooperative International Gas R&D", *Proceedings,* 17th World Gas Conference, International Gas Union, June 1988.

24. Barnard V.J. et. al.; *CNG Facts Folder,* Ministry of Energy, New Zealand, 1988.

25. Bartham J.; "Brazil Reduces Fuel Alcohol Subsidy", *Financial Times,* 24th June 1988.

26. Bertilsson B., et. al. (Volvo Truck Corporation); *Experience of Heavy-Duty Alcohol Fuelled Diesel-Ignition Engines,* 1988.

27. Birrell J.S. et. al. (Repco Engine Technical Centre); *Ethyl Alcohol as a Fuel or Fuel Supplement in Otto or Diesel Cycle Engines in Australia: An Evaluation of the Suitability - Section I: Spark Ignition Investigation,* National Energy R&D and Demonstration Program, Report No. 352, Department of Resources and Energy, Australia, 1984.

28. Bleviss D.; *Preparing for the Energy Challenges of the 1990s, Energy Efficiency: The Key to Reducing the Vulnerability of the Nation's Transportation Sector,* International Institute for Energy Conservation, United Kingdom, 1988.

29. Braun A.R.; *Automotive Fuel Extenders from C4 Hydrocarbons,* National Energy R&D and Demonstration Program, Report No. 192, Department of Resources and Energy, Australia, June 1983.

30. British Petroleum; *Statistical Review of World Energy,* July 1989.

31. Butler S.; "Shell and BP Ready to Produce Transport Fuel from Natural Gas", *Financial Times,* 2nd August 1988.

32. California Energy Commission; *California's Methanol Fleet Program,* 1985.

33. California Energy Commission; *California's Methanol Program, Evaluation Report, Volume 1 Executive Summary,* November 1986.

34. California Energy Commission, California Air Resources Board, and South Coast Air Quality Management District; *Light-Duty Methanol Vehicle Workshop,* Los Angeles, April 1987.

35. Campbell S. (United States Department of Energy); "Assessment of the Costs and Benefits of Flexible and Alternative Fuel Use in the United States Transportation System", *Testimony,* United States House of Representatives, 1987.

36. Canadian Department of Energy Mines and Resources, Office of Energy R&D; *Plans and Programs of Energy Research and Development of the Interdepartmental Panel on Energy Research and Development of the Government of Canada,* 1987.

37. Canadian National Energy Board; *Canadian Energy Supply and Demand 1985-2005,* 1986.

38. Cedigaz (France); *Natural Gas in the World in 1988,* Paris, 1989.

39. Chem Systems Inc.; "Methanol Supply/Demand for the United States and the Impact of the Use of Methanol as a Transportation Fuel", *Gas Energy Review,* American Gas Association, 1987.

40. *Clean-Coal/Synfuels Letter;* various issues, 1987 and 1988.

41. Clemmens W.; *Performance of Sequential Port Fuel Injection on a High Compression Ratio Neat Methanol Engine,* SAE Technical Paper No. 872070, 1987.

42. Cohen L.H. and Muller H.L.; "Technology, Methanol Cannot Economically Dislodge Gasoline", *Oil and Gas Journal,* 28th January 1985.

43. Consiglio Nazionale delle Richerche; *Use of Methanol for Motor Vehicles,* 1982.

44. Conisglio Nazionale delle Richerche; *Ethanol via Fermentation - Production and Use in Vehicles,* 1979.

45. De Marco V.R.; "UMTA Methanol Programme", *Proceedings,* On the Road with Natural Gas Conference, 1987.

46. DeLuchi M.A.; *Transportation Fuels and the Greenhouse Effect,* Division of Environmental Studies, University of California, Davis, 1987.

47. DeLuchi M.A. et. al.; *A Comparative Analysis of Future Transportation Fuels,* Research Report UCB-ITS-RR-87-13, Institute of Transportation Studies, University of California, 1987.

48. Difiglio C. et. al.; "Economic and Security Issues of Methanol Supply", *Proceedings,* SAE Fuels and Lubricants Meeting, 1987.

49. Dyne, P.J. (Canadian Department of Energy, Mines and Resources); "The Past and Future of R&D in Alternative Transportation Fuels", *Proceedings,* Fourth Windsor Workshop on Alternative Fuels, Canada, 20th-22nd June 1988.

50. Ecklund, E.E.; "Status of Commercialization of Alternative Fuels for Highway Vehicles", *Proceedings,* Energy Technology '87 Conference, 1987.

51. Ecklund, E.E.; "Options for and Recent Trends in Use of Alternative Transportation Fuels", *Proceedings,* United Nations Meeting on Energy in Human Settlements, 1986.

52. Ecklund E.E. and Douthit W.H. (editors); *Proceedings of SAE Conference on Transportation Fuel Alternatives for North America into the 21st Century,* 1985.

53. Egebäck K.E. et. al.; *Chemical and Biological Characterization of Exhaust Emissions from Vehicles Fueled with Gasoline, Alcohol, LPG, and Diesel,* 1983.

54. Egebäck K.E. et. al.; *Regulated and Unregulated Pollutants from Vehicles Fueled with Alcohols,* November 1988.

55. Energy Research, Development and Information Centre, University of New South Wales; *Proceedings,* Workshop on Diesel Substitution by Gas in Vehicles, Australia, October 1987.

56. Energy and Environmental Analysis, Inc.; *Distribution of Methanol for Motor Vehicle Use in the California South Coast Air Basin,* Contract No. 68-03-1865, U.S. Environmental Protection Agency, 1986.

57. Energy and Environmental Analysis, Inc.; *Documentation of the Characteristics of Technological Improvements Utilized in the Technology/Cost Segment Model,* 1985.

58. Ente Nazionale Idrocarburi; *Ethanol for Motor Fuel as an Outlet for EEC Wheat Surplus,* Rome, September 1985.

59. Federal Republic of Germany, Bundesministerium für Forschung und Technologie; *Alternative Energy Sources for Road Transport - Methanol,* 1984.

60. Ford Motor Company; "Presentation on the Flexible Fuel Vehicle to IEA Thematic Review Team", 1987.

61. Francis R.J. et. al. (Energy Technology Support Unit, Harwell); *Prospects for Improved Fuel Economy and Fuel Flexibility in Road Vehicles,* Department of Energy, United Kingdom, 1988.

62. Gardiner D.P.; *Spark-Ignition Cold Starting with Methanol-Based Fuel Blends,* SAE Technical Paper No. 872067, 1987.

63. Garibaldi P. (Ente Nazionale Idrocarburi); Automotive Synfuels: *A Possible Strategy for the Future,* Rome, 1985.

64. Garibaldi P.; *Environmental Problems from Synfuels Production and Utilization,* Rome, 1985.

65. Garibaldi P.; "Fuel Oxygenates: the Right Solutions for Cleaner Cars and a Better Environment", *Proceedings,* European Fuel Oxygenates Second Conference, 1987.

66. Garibaldi P.; "Greenhouse Effect of Alcohol Fuels", presentation at IEA Alcohol Fuels Implementing Agreement Meeting, Kiruna, Sweden, June 1988.

67. Garibaldi P. et. al.; "What's Next", *Proceedings,* European Fuel Oxygenates Conference, 1986.

68. Garibaldi P. et. al.; "The Developing Oxygenates Market in Southern Europe", *Proceedings,* European Octane and Fuel Oxygenates Conference, Geneva, May 5-6th, 1987, GEIR-36/87, April 1987.

69. *Gas Matters,* 27th April 1989 and 30th June 1989.

70. Gas Research Institute; *Natural Gas Vehicles: The International Experience,* Issue Brief 1988-9, 13th May, 1988.

71. Gold M.D. and Moulis C.E.; *Emission Factor Data Base for Prototype Light-Duty Methanol Vehicles,* SAE Technical Paper No. 872055, 1987.

72. Government of Alberta, *Alberta Oil Supply, 1985-2010,* Calgary, Canada, 1985.

73. Hallett P. et. al.; "An Assessment Methodology for Alternative Fuels Technologies", *Transportation Research Record No. 1092,* 1986.

74. Higgs N.B. (Australian Gas Association); "Natural Gas for Vehicles", address to the National Energy Research, Development and Demonstration Council, 1986.

75. Hirao, O. and Pefley K. (Eds); *Present and Future Automotive Fuels: Performance and Exhaust Emissions,* John Wiley and Sons, New York, 1987.

76. Holcomb C.M. et.al. (Oak Ridge National Laboratory); *Transportation Energy Data Book: Edition 9,* Office of Transportation Systems, U.S. Department of Energy, 1987.

77. Information Resources, Inc. (IRI) and Octane Week; "Air Quality Issues: Changing America's Motor Fuel Business", *1988 Executive Seminar,* January 1988.

78. Iwai N. et. al.; *A Study on Cold Startability and Mixture Formation of High-Percentage Methanol Blends,* SAE Technical Paper No. 880044, 1988.

79. Jack Faucett & Associates; *Methanol Prices During Transition, Final Report,* United States Environmental Protection Agency, 1987.

80. Langley K.F. (Energy Technology Support Unit, Harwell); *A Ranking of Synthetic Fuel Options for Road Transport Applications in the United Kingdom,* Paper R-33, 1987.

81. Larsen R.P. et. al.; "Rationale for Converting the United States Transportation System to Methanol Fuel", *Proceedings of the 7th International Symposium on Alcohol Fuels,* 1986.

82. Lipari F. and Colden F.L.; *Aldehyde and Unburned Fuel Emissions from Developmental Methanol-Fueled 2.5L Vehicles,* SAE Technical Paper No. 872051, 1987.

83. Machiele P.A.; *Flammability and Toxicity Tradeoffs with Methanol Fuels,* SAE Technical Paper No. 872064, 1987.

84. McNutt B. et. al.; *The Cost of Making Methanol Available to a National Market,* SAE, 1987.

85. McNutt B. and Ecklund E.E.; *Is there a Government Role in Methanol Market Development?* SAE, 1986.

86. MacRae K.M. (Canadian Energy Research Institute); *An Assessment of the Potential for Coal-Derived Syncrudes in Canada,* Study No. 27, 1988.

87. Maiden C.; "The New Zealand Gas to Gasoline Project", *Proceedings,* Methane Conversion Symposium, 1987.

88. Manne A.S.; *A Progress Report on the Formulation of AFTM: An Alternative Fuels Trade Model,* Stanford University, United States, 1987.

89. Markus H.; "Projektleitung Biologie Ökologie und Energie", *Proceedings,* Ecofuel Workshop, Milan, 8th March 1988.

90. Marrow J.E. et. al. (Energy Technology Support Unit, Harwell); *An Assessment of Bio-Ethanol as a Transport Fuel in the United Kingdom,* Paper R-44, 1987.

91. Moody K. (ICI Australia Operations Pty. Ltd.); *Additives for Methanol for its Use as Diesel Fuel, Reporting Period 1.7.83 - 31.12.83,* National Energy R&D and Demonstration Program Progress Report No. 82/2103, Department of Resources and Energy, Australia, 1983.

92. Moody K. (ICI Australia Operations Pty. Ltd.); *Additives for Methanol for its Use as Diesel Fuel,* Commonwealth of Australia, National Energy Research, Development and Demonstration Program, End of Grant Report Number 510, Department of Resources and Energy, 1985.

93. Morrison, Cooper & Partners New Zealand; *Methanol Distribution - Options & Costs,* Final Report, Liquid Fuels Trust Board, Contract 730/01/4, 1985.

94. Moses D. et. al.; *A Review of Methanol Vehicles and Air Quality Impacts,* SAE Technical Paper No. 872053, 1987.

95. Mueller Associates Inc.; *Alcohols from Biomass: State-of-Knowledge Survey of Environmental, Health and Safety Aspects,* Office of Environmental Programs, United States Department of Energy, 1981.

96. National Advisory Panel on Cost-Effectiveness of Fuel Ethanol Production; *Fuel Ethanol Cost-Effectiveness Study,* Final Report, 1987.

97. New Energy Development Organization (Japan); "Alcohol and Biomass Energy Technology", *Annual Report,* 1986.

98. New Energy Development Organization (Japan); *Data Collection and Surveys on Diesel-Substitute Methanol Vehicle through Prototype Vehicle Driving Test and Other Studies,* 1987.

99. New Energy Development Organization (Japan); *Data Collection and Surveys on Diesel-Substitute Methanol Vehicle through Prototype Vehicle Driving Test and Other Studies,* 1988.

100. New Energy Development Organization (Japan); Nuclear Research Centre (Karlsruhe, Germany); *Technology Assessment of Various Coal Fuel-Options,* 1987.

101. O'Hare T.E. et. al.; *Methanol for Transportation of Natural Gas Values,* Brookhaven National Laboratory, United States, 1987.

102. *Oil & Gas Journal;* various issues, 1987, 1988 and 1989.

103. Okkenn P.A. (Energy Study Centre); *Impacts of Environmental Constraints on Energy Technology,* The Case of NOx in the Netherlands, 1988.

104. Patel K.S. et. al.; *The Performance Characteristics of Indolene-MPHA Blends in a Spark-Ignition Engine,* SAE Technical Paper No., 872068, 1987.

105. Pereira A.; *Ethanol Employment and Development: Lessons from Brazil,* International Labour Office, Geneva, 1986.

106. Petroleum Energy Center; *Feasibility Study on the Utilization of Neat Methanol (M80-M100) Fuel for the Automobile in FY1985,* Vols. 1 and 2, 1986.

107. Petroleum Energy Center; *Feasibility Study on the Utilization of Methanol for the Automobile in FY1986 (Outline of Study Results),* 1987.

108. Philp R.J.; *Methanol Production from Biomass,* National Research Council, Canada, 1986.

109. Quadflieg H.; *Methanol Fuels, R&D Activities in the Federal Republic of Germany,* 1987.

110. Reeves R.R. and Lom E.J.; *Ethanol/Distillate Emulsified Blends in Diesel Engines,* National Energy R&D and Demonstration Program, Report No. 357, Department of Resources and Energy, Australia, 1984.

111. Research Association for Petroleum Alternatives Development; *Research and Development on Synfuels,* Annual Technical Report, 1987.

112. SAE, *Proceedings,* International Fuels and Lubricants Meeting, Toronto, Canada, 1987.

113. Santini D.J.; *The Past and Future of the Petroleum Problem: The Increasing Need to Develop Alternative Transportation Fuels,* Argonne National Laboratory, United States, 1987.

114. Schiefelbein G.F. et. al.; *Biomass Thermochemical Conversion Program 1986 Annual Report,* U.S. Department of Energy, 1987.

115. Sell D.P.M.; *National Economics of CNG, The Planning Implications of the Current CNG Programme and the Desirability and Achievability of Longer Term Alternatives,* Ministry of Energy, New Zealand, 1987.

116. Shell International Petroleum; *The Shell Middle Distillate Synthesis Process,* 1985.

117. Siewert R.M. and Groff E.G.; *Unassisted Cold Starts to -29 Degrees Centigrade and Steady-State Tests of a Direct-Injection Stratified-Charge (DISC) Engine Operated on Neat Alcohols,* SAE Technical Paper No. 872066, 1987.

118. Sperling D.; *New Transportation Fuels: A Strategic Approach to Technological Change,* University of California Press, United States, 1988.

119. Stone and Webster Engineering Corporation; *Economic Feasibility Study of a Wood Gasification Based Methanol Plant,* Boston, United States, 1985.

120. Swedish Government Committee on Automotive Air Pollution; *Motor Vehicles and Cleaner Air,* 1986.

121. Swedish Ministry of Environment and Energy; *Trends - Alcohols as Vehicle Fuels,* 1986.

122. Swedish Motor Fuel Technology Co.; *Trends - Alcohols as Motor Fuels,* various issues, 1987 and 1988.

123. Swedish National Board for Technical Development; *Project M 100 - A Test with Methanol-Fueled Vehicles in Sweden,* 1987.

124. Sypher:Mueller International Inc.; *Milestone, A Report on the Status of Project MILE -Methanol in Large Engines,* 1988.

125. Sypher:Mueller International Inc.; *Project MILE 1986 Annual Report,* Canada, 1986.

126. Sypher:Mueller International Inc.; Study Papers for the IEA Executive Committee on Alcohols and Alcohol Blends as Motor Fuels, 8th Executive Committee Meeting, 1988.

127. Tanner J. (CSR Chemicals Ltd.); *Enhanced Extension of Petrol With Aqueous Alcohol,* National Energy R&D and Demonstration Program, Report No. 201, Department of Resources and Energy, Australia, 1983.

128. Tanner J. (CSR Chemicals Ltd.); *Enhanced Extension of Petrol with Aqueous Alcohol,* National Energy R&D and Demonstration Program, Report No. 310, Department of Resources and Energy, Australia, 1984.

129. Task Force on Alternative Fuels; *Alternative Transportation Fuels: Review of Research Activity in Canada,* 1987.

130. Tsuruno Y.; *Ministry of Transport's Fleet Test of Methanol Vehicles,* 1988.

131. United States Department of Agriculture; *Ethanol: Economic and Policy Trade-offs,* 1988.

132. United States Department of Commerce, International Trade Administration; *A Competitive Assessment of the U.S. Methanol Industry,* 1985.

133. United States Department of Energy; *Assessment of Costs and Benefits of Flexible and Alternative Fuel Use in the United States Transportation Sector, Progress Report One: Context and Analytical Framework,* 1988.

134. United States Department of Energy; *Assessment of Methane-related Fuels for Automotive Fleet Vehicles,* 1982.

135. United States Department of Energy; *Project Planning Document, Highway Vehicle Alternative Fuels Utilization Program (AFUP),* Assistant Secretary for Conservation and Renewable Energy, 1985.

136. United States Department of Energy; *National Methanol Fuel Market Development,* 1988.

137. United States Environmental Protection Agency, Office of Mobile Sources; *Air Quality Benefits of Alternative Fuels,* Vice-President's Task Force on Alternative Fuels, 1987.

138. United States Environmental Protection Agency; *Guidance on Estimating Motor Vehicle Emission Reduction from the Use of Alternate Fuels and Fuel Blends - Final Draft for Task Force Review,* 1988.

139. United States General Accounting Office; "Alternative Fuels, Status of Methanol Vehicle Development", Briefing Report to the Chairman, Subcommittee on Fossil and Synthetic Fuels, Committee on Energy and Commerce, House of Representatives, 1986.

140. United States Office of Technology Assessment; *Increased Automobile Fuel Efficiency and Synthetic Fuels,* 1982.

141. von Hippel F.; "Automobile Fuel Economy", *Energy,* Vol. 12, 1987.

142. Wagner T.O. et. al. (Amoco); *Comparative Economics of Methanol and Gasoline,* SAE Technical Paper No. 872061, 1987.

143. Walde N. et. al.; *A Study of the Organic Emission from a Turbocharged Diesel Engine Running on 12 Per Cent Hexyl Nitrate Dissolved in Ethanol,* SAE Technical Paper No. 840367, 1984.

144. Wan E.I. (Science Applications, Inc.); *Economics of Methanol Production from Indigeneous Resources,* 1983.

145. Wan E.I. and Price J.D. (Science Applications, Inc.); *Technical and Economic Assessment of Liquid Fuel Production from Biomass, Final Research Report,* U.S. Department of Energy, 1982.

146. Weide J. et. al.; "Vehicle Operation with Variable Methanol/Gasoline Mixtures", *Proceedings,* Sixth International Symposium on Alcohol Fuels Technology, 1984.

147. World Methanol Conference; *Proceedings,* December 1987.

148. Wright J.D; *Ethanol from Lignocellulose-An Overview,* Solar Energy Research Institute, United States, 1987.

149. Wright J.D. and Power A.J.; *Comparative Technical Evaluation of Acid Hydrolysis Processes for Conversion of Cellulose to Alcohol,* Solar Energy Research Institute, United States, 1988.

150. Wubs K. and Deckers R.; *Strategic PEO-Study into Alternative Fuels for Internal Combustion Engines,* Projektbeheerbureau Energieonderzoek (PEO), The Netherlands, 1988.

Reference numbers refer to both notes and sources listed below. General assumptions:
Methanol to gasoline (premium unleaded) equivalent system conversion factor = 2.0
Ethanol to gasoline equivalent system conversion factor = 1.5
Exchange rates: 1987 average for OECD countries; other years deflated to 1987 values.

Notes for Table A:

1.	CNG - gas price:	1987 A$2-6 per GJ
	CNG - compression and dispensing cost: (assumes no additional pipeline cost)	1987 A$0.05-0.10 per litre gasoline equivalent
	Ethanol - sugar cane feedstock price:	1987 A$20-25 per ton
2.	VHO - feedstock price:	1984 $4.1 per GJ
	Methanol, synthetic gasoline and synthetic diesel fuel - gas feedstock price:	1984 $3.70 per GJ
	Ethanol - wheat feedstock price: (Discount rate 10 percent)	1986 £112 per ton
3.	Methanol - gas feedstock price:	1985 $4 per MBTU
4.	VHO - raw oil feedstock price:	1987 $12.4 per barrel
	Gasoline from advanced Mobil process with alkylation - gas feedstock price:	1987 $2 per MBTU
5.	Methanol and synthetic gasoline - gas feedstock price:	1983 DM5 per GJ
6.	Methanol - gas feedstock price: (internal rate of return ranges from 10-30 percent)	1987 $2 per MBTU
7.	Methanol - U.S.-sourced gas feedstock price:	1987 US $2 per MBTU
	Remote-sourced gas feedstock price:	1987 US $0.75 per MBTU
8.	Methanol - lower end of range refers to 500 t/d plant, higher end refers to 100 t/d plant - wood feedstock price:	1987 US $30 per ton

10. Lower end of cost range uses lignite, higher end uses hard coal.

Sources for Table A:

1. Australian Minerals and Energy Council; 1987 (Ref. 22).
2. Langley K.F., ETSU, United Kingdom; 1987 (Ref. 80) and J. Marrow (ETSU).
3. United States Department of Energy; 1988 (Ref. 133).
4. Ministry of International Trade and Industry, Japan, 1987.
5. Markus H., Federal Republic of Germany; 1988 (Ref. 89).
6. World Methanol Conference; Proceedings, December 1987 (Ref. 147).
7. Wagner, T.O. et al, Amoco; 1987 (Ref. 142).
8. Stone and Webster Engineering Corporation; 1985 (Ref. 119).
9. Canadian National Engineering Board; 1986 (Ref. 37).
10. Nuclear Research Centre, Karlsruhe, Federal Republic of Germany; 1987 (Ref. 100).

ANNEX B

BACKGROUND ON VEHICLE EMISSION CONTROL REGULATIONS

Table B-1 shows current light-duty vehicle exhaust emission limits for the United States, Japan and the European Community, and estimated actual emissions of the 1987 United States and EC car "parcs". Light duty vehicles are defined as gasoline-fuelled vehicles able to carry 12 passengers.

Table B-1: **Vehicle Exhaust Emission Limits and Actual Emissions**
(grams per kilometer)

	CO	**HC**	**NO_x**	**Comment**
US[1]-Limits	2.1	0.26	0.63	Requires 3-way catalyst.
US-Actual	6.9	0.7	1.2	
Japan-Limits	2.7	0.39	0.48	
EEC-Limits	14.3-27.1	3.0-6.9	1.7-3.1	Depends on vehicleweight and method of calculation.
EEC-Actual	17.5-37.6	4.1-5.9	2.3-3.4	This is a general point, which makes comparisons difficult.

1. Applies also to Canada, but excludes California.

Source: P. Garibaldi (Ref. 65); MITI (Japan)

The present EEC situation will change to a significant extent with the enforcement of the recently approved directive amendment that rules as shown in Table B-2. The larger cars will require three-way catalyst technology. Probably for the medium size cars, an oxidizing catalyst would be enough; for smaller size cars a lean combustion engine should be able to meet the standards.

Table B-2: **EEC Vehicle Emissions Regulations**

Engine Displacement	Enforcement Date		Pollutant Limits (grams per kilometer)		
	New Models	New Registrations	CO	HC+NOx	NOx
Greater than 2.0 litres	Oct. 1988	Oct. 1989	6.2	1.6	0.86
1.4 - 2.0 litres	Oct. 1991	Oct. 1993	7.4	2.0	N.D.
Less than 1.4 litres	Oct. 1992	Oct. 1991	11.1	3.7	1.48

Full penetration of the new limits will require about twelve years, based on the average ratio in the EEC between new cars and existing car population. Consequently, it will only be in the first decade of the next century that the EEC situation will reach the present one in the United States.

A ruling on evaporative emissions was first enforced in the United States on the 1971 model year, limited to the light duty vehicles. It was later (1978) modified in the test procedure and limits and it was applied also to the heavy duty vehicles operating on gasoline. Table B-3 summarises the United States regulations through the years. There is at present no ruling in the EEC for evaporative emissions although this problem is presently under consideration.

Table B-3: **Evaporative Emissions**
(grams per test)

Years	Light Duty Vehicles
1971-1977[1]	2.0
1978-1980[2]	6.0
1981-1987[3]	2.0

1. Carbon canister test method.
2. The method was changed to "Sealed Housing Emission Determination" (SHED) by which emissions from the whole vehicle are measured.
3. As from 1987, also for Canada.

There is also increasing concern regarding emissions of diesel particulate matter, which have mutagenic activity. Particulate emission from Diesel engines is generally about one or two orders of magnitude higher than from gasoline engines. Accordingly, Canada and the United States have established the stringent particulate emission standards set out in Table B-4. Sweden also proposes to bring in standards controlling diesel particulate emissions, and the European Community is considering the issue.

Table B-4: **Pending Diesel Engine Emissions Limits United States**
(grams per brake-horsepower-hour)

Year	HC	CO	NO_x	Particulates
1988[1]	1.3	15.5	6.0	0.6
1991-1993	1.3	15.5	5.0	0.25 0.1[2]
1994 on	1.3	15.5	5.0	0.1[3]

1. Canada, as from 1 December 1988; already in existence for California.
2. Applicable to urban bus engines.
3. All engines.

Source: United States Environmental Protection Agency.

Other environmental issues include the effects of episodes of photochemical smog and long-term build-up of background ozone concentrations (both from the reaction of volatile organic compounds and nitrogen oxides); emission of toxic pollutants, including benzene, aldehydes, polyaromatic hydrocarbons and dioxins; and the broader "greenhouse" pollutants, including CO_2, N_2O, and methane.

ANNEX C

GLOSSARY

Adsorption: Process whereby gas may be held on the surface of a solid by physiochemical forces.

AGA: American Gas Association

Alcohol fuel; fuel methanol; fuel ethanol: Fuel containing a minimum of 85 percent alcohol/methanol/ethanol by volume (for example, M85, M100).

Aromatics: Compounds containing a benzene ring.

Benzene: (C_6H_6): A colourless and highly inflammable liquid. The simplest member of the aromatic series of hydrocarbons. Used as a component of petrol.

Biomass: Plant and animal matter - wastes and residues that may be used as a source of useable energy and/or chemicals.

CAFE: Corporate Average Fuel Economy (United States)

Carbon dioxide (CO_2): One of the so-called "greenhouse gases", the accumulation of which in the atmosphere may lead to global climate change.

Carbon monoxide (CO): A photochemical smog precursor.

Cetane number: A measure of the ignition quality of Diesel engine fuels, based on cetane ($C_{16}H_{34}$).

CNG: Compressed natural gas (methane)

CO_2: Carbon dioxide

CO: Carbon monoxide

Dedicated vehicle: Vehicle operating on a single fuel only, for example, compressed natural gas.

Diesel engine: A compression-ignition engine, at present using diesel fuel. However, for use with alcohol fuels, spark-ignition assistance can be required.

Ethanol (C_2H_5OH): The next simplest alcohol after methanol.

DOE: Department of Energy (United States)

EPA: Environmental Protection Agency (United States)

Flexible fuel vehicle (FFV): One which is able, by the use of sensors, to operate on mixtures in any proportion of two fuels used in the same tank, for example, methanol and gasoline.

Formaldehydes: Chemical compounds produced by oxidation of alcohols, for example, during combustion in engines.

HC: Hydrocarbons

Hydrolysis: Chemical decomposition involving the formation of water.

IEA: International Energy Agency

Lean: Fuel/air mixture with less fuel (more air) than stoichiometric.

Lignocellulosics: Substances composed of lignin and cellulose, for example, wood.

LNG: Liquefied natural gas (methane)

LPG: Liquefied petroleum gas (propane plus butane)

Methanol (CH_3OH): The simplest alcohol.

Motor octane number (MON): The octane number of a motor fuel determined in a single cylinder laboratory test engine, and under *high* "engine-severity" conditions. Gives an indication of the *high*-speed knock properties of the fuel.

Mutagenic: The property of causing cell mutation, which may lead to cancer.

NO_x (Oxides of nitrogen: N_2O and NO_2): involved in the formation of photochemical smog.

Octane number: An indication of the anti-knock quality of a motor fuel. The percentage by volume of iso-octane in a mixture of iso-octane and normal heptane which has the same knocking characteristics as the motor fuel under test.

Otto engine: Spark-ignition engine, at present, primarily using gasoline.

Oxygenate: An organic compound containing oxygen and having properties as a fuel which are compatible with petrol. Includes alcohols and ethers.

Particulates: Finely divided solids emitted from vehicle exhausts, perceived as smoke.

Reid vapour pressure: The pressure exerted by a vapour in contact with its liquid, as measured under standard conditions; gives some indication of the volatility of a liquid.

Research octane number (RON): The octane number of a motor fuel determined in a single cylinder laboratory test engine, and under *mild* "engine-severity" conditions. Gives an indication of the *low*-speed knock properties of the fuel.

Stoichiometric: Descriptive of a reaction mixture in which the reactants are present in the exact amounts necessary for the reaction to go to completion with no reactants being left.

Synthesis gas: Mixture of carbon monoxide (CO) and hydrogen (H_2), made by reacting gas or coal with steam and oxygen, and used to make liquid fuels such as methanol.

Very heavy oils (VHO): For the purposes of this study, construed to be oils of gravity less than 10° API, including bitumens and tar sands.

NO_x: Oxides of nitrogen, NO and NO_2, involved in the formation of photochemical smog.

Octane number: An indication of the anti-knock quality of a motor fuel. The percentage by volume of iso-octane in a mixture of iso-octane and normal heptane which has the same knocking characteristics as the motor fuel under test.

Otto engine: Spark-ignition engine, at present primarily using gasoline.

Oxygenates: Any organic compound containing oxygen and thus able to operate as a fuel which is more compatible with petrol. Includes alcohols and ethers.

Particulates: Finely divided solids emitted from vehicle exhausts; perceived as smoke.

Reid vapour pressure: The pressure exerted by a vapour in contact with its liquid, as measured under standard conditions; gives some indication of the volatility of a liquid.

Research octane number (RON): The octane number of a motor fuel determined in a single-cylinder laboratory test engine, and under mild "engine severity" conditions. Gives an indication of the low-speed knock properties of the fuel.

Stoichiometric: Description of a reaction mixture in which the reactants are present in the exact amounts necessary for the reaction to go to completion with no reactants left over.

Synthesis gas: Mixture of carbon monoxide (CO) and hydrogen (H_2) made by reacting gas or coal with steam and oxygen; can be used to make liquid fuels such as methanol.

Very heavy oils (VHO): For the purposes of this study, considered to be oils of gravity less than 10° API, including bitumens and tar sands.

ANNEX D

MEMBERS OF THE EXPERT GROUP ON SUBSTITUTE FUELS FOR ROAD TRANSPORT

AUSTRALIA
Mr. W. G. McGregor
Department of Primary Industries and Energy

CANADA
Mr. P. Reilly-Roe
Department of Energy, Mines and Resources

GERMANY
Unit
Dr. F. Conrad
Kernforschungszentrum

Dr.-Ing. H. Quadflieg
Technischer Uberwachungs-Verein

ITALY
Dr. P.P. Garibaldi
Ecofuel S.p.a.
ENI

JAPAN
Mr. S. Nakayama and
Mr. K. Sugimoto
New Energy and Industrial Technology Development

NETHERLANDS
Mr. K. Kouma
Netherlands Agency for Energy and the Environment

NEW ZEALAND
Dr. Richardson
Ministry of Energy

NORWAY
Mr. J. Hoiland
Royal Norwegian Council for Scientific and Industrial Research

SWEDEN
Mr. U. Backlund
Volvo Personvagnar

Mr. G. Kinbom
Swedish National Board for
Technical Development

UNITED KINGDOM
Mr. J. Marrow
Energy Technology Support

UNITED STATES
Dr. F. W. Bowditch
Motor Vehicles Manufacturers
Association

Mr. C. Difiglio
Department of Energy

Mr. E. E. Ecklund
Department of Energy

Mr. G. F. Keller
Engine Manufacturers
Association

Mr. R. Moorer
Department of Energy

COMMISSION OF THE
EUROPEAN COMMUNITIES
Mr. M. Roma
Directorate General XVII

IEA SECRETARIAT
Dr. S. Garribba
Mr. L. Boxer
Mr. J. Skeer
Dr. R. Stuart
Dr. M. Taylor

WHERE TO OBTAIN OECD PUBLICATIONS
OÙ OBTENIR LES PUBLICATIONS DE L'OCDE

Argentina - Argentine
Carlos Hirsch S.R.L.
Galeria Güemes, Florida 165, 4° Piso
1333 Buenos Aires
Tel. 30.7122, 331.1787 y 331.2391
Telegram: Hirsch-Baires
Telex: 21112 UAPE-AR. Ref. s/2901
Telefax:(1)331-1787

Australia - Australie
D.A. Book (Aust.) Pty. Ltd.
11-13 Station Street (P.O. Box 163)
Mitcham, Vic. 3132 Tel. (03)873.4411
Telex: AA37911 DA BOOK
Telefax: (03)873.5679

Austria - Autriche
OECD Publications and Information Centre
4 Simrockstrasse
5300 Bonn (Germany) Tel. (0228)21.60.45
Telex: 8 86300 Bonn
Telefax: (0228)26.11.04

Gerold & Co.
Graben 31
Wien I Tel. (0222)533.50.14

Belgium - Belgique
Jean De Lannoy
Avenue du Roi 202
B-1060 Bruxelles
Tel. (02)538.51.69/538.08.41
Telex: 63220 Telefax: (02)538.08.41

Canada
Renouf Publishing Company Ltd.
1294 Algoma Road
Ottawa, Ont. K1B 3W8 Tel. (613)741.4333
Telex: 053-4783 Telefax: (613)741.5439
Stores:
61 Sparks Street
Ottawa, Ont. K1P 5R1 Tel. (613)238.8985
211 Yonge Street
Toronto, Ont. M5B 1M4 Tel. (416)363.3171

Federal Publications
165 University Avenue
Toronto, ON M5H 3B9 Tel. (416)581.1552
Telefax: (416)581.1743
Les Publications Fédérales
1185 rue de l'Université
Montréal, PQ H3B 1R7 Tel.(514)954-1633

Les Éditions La Liberté Inc.
3020 Chemin Sainte-Foy
Sainte-Foy, P.Q. G1X 3V6
Tel. (418)658.3763
Telefax: (418)658.3763

Denmark - Danemark
Munksgaard Export and Subscription Service
35, Norre Sogade, P.O. Box 2148
DK-1016 Kobenhavn K
Tel. (45 33)12.85.70
Telex: 19431 MUNKS DK
Telefax: (45 33)12.93.87

Finland - Finlande
Akateeminen Kirjakauppa
Keskuskatu 1, P.O. Box 128
00100 Helsinki Tel. (358 0)12141
Telex: 125080 Telefax: (358 0)121.4441

France
OECD/OCDE
Mail Orders/Commandes par correspondance:
2 rue André-Pascal
75775 Paris Cedex 16 Tel. (1)45.24.82.00
Bookshop/Librairie:
33, rue Octave-Feuillet
75016 Paris Tel. (1)45.24.81.67
(1)45.24.81.81
Telex: 620 160 OCDE
Telefax: (33-1)45.24.85.00

Librairie de l'Université
12a, rue Nazareth
13602 Aix-en-Provence Tel. 42.26.18.08

Germany - Allemagne
OECD Publications and Information Centre
4 Simrockstrasse
5300 Bonn Tel. (0228)21.60.45
Telex: 8 86300 Bonn
Telefax: (0228)26.11.04

Greece - Grèce
Librairie Kauffmann
28 rue du Stade
105 64 Athens Tel. 322.21.60
Telex: 218187 LIKA Gr

Hong Kong
Government Information Services
Publications (Sales) Office
Information Service Department
No. 1 Battery Path
Central Tel. (5)23.31.91
Telex: 802.61190

Iceland - Islande
Mal Mog Menning
Laugavegi 18, Postholf 392
121 Reykjavik Tel. 15199/24240

India - Inde
Oxford Book and Stationery Co.
Scindia House
New Delhi 110001 Tel. 331.5896/5308
Telex: 31 61990 AM IN
Telefax: (11)332.5993
17 Park Street
Calcutta 700016 Tel. 240832

Indonesia - Indonésie
Pdii-Lipi
P.O. Box 269/JKSMG/88
Jakarta12790 Tel. 583467
Telex: 62 875

Ireland - Irlande
TDC Publishers - Library Suppliers
12 North Frederick Street
Dublin 1 Tel. 744835/749677
Telex: 33530 TDCP EI Telefax : 748416

Italy - Italie
Libreria Commissionaria Sansoni
Via Benedetto Fortini, 120/10
Casella Post. 552
50125 Firenze Tel. (055)645415
Telex: 570466 Telefax: (39.55)641257
Via Bartolini 29
20155 Milano Tel. 365083
La diffusione delle pubblicazioni OCSE viene assicurata dalle principali librerie ed anche da:
Editrice e Libreria Herder
Piazza Montecitorio 120
00186 Roma Tel. 679.4628
Telex: NATEL I 621427

Libreria Hoepli
Via Hoepli 5
20121 Milano Tel. 865446
Telex: 31.33.95 Telefax: (39.2)805.2886

Libreria Scientifica
Dott. Lucio de Biasio "Aeiou"
Via Meravigli 16
20123 Milano Tel. 807679
Telefax: 800175

Japan- Japon
OECD Publications and Information Centre
Landic Akasaka Building
2-3-4 Akasaka, Minato-ku
Tokyo 107 Tel. 586.2016
Telefax: (81.3)584.7929

Korea - Corée
Kyobo Book Centre Co. Ltd.
P.O. Box 1658, Kwang Hwa Moon
Seoul Tel. (REP)730.78.91
Telefax: 735.0030

Malaysia/Singapore - Malaisie/Singapour
University of Malaya Co-operative Bookshop Ltd.
P.O. Box 1127, Jalan Pantai Baru 59100
Kuala Lumpur
Malaysia Tel. 756.5000/756.5425
Telefax: 757.3661

Information Publications Pte. Ltd.
Pei-Fu Industrial Building
24 New Industrial Road No. 02-06
Singapore 1953 Tel. 283.1786/283.1798
Telefax: 284.8875

Netherlands - Pays-Bas
SDU Uitgeverij
Christoffel Plantijnstraat 2
Postbus 20014
2500 EA's-Gravenhage Tel. (070)78.99.11
Voor bestellingen: Tel. (070)78.98.80
Telex: 32486 stdru Telefax: (070)47.63.51

New Zealand -Nouvelle-Zélande
Government Printing Office
Customer Services
P.O. Box 12-411
Freepost 10-050
Thorndon, Wellington
Tel. 0800 733-406 Telefax: 04 499-1733

Norway - Norvège
Narvesen Info Center - NIC
Bertrand Narvesens vei 2
P.O. Box 6125 Etterstad
0602 Oslo 6
Tel. (02)67.83.10/(02)68.40.20
Telex: 79668 NIC N Telefax: (47 2)68.53.47

Pakistan
Mirza Book Agency
65 Shahrah Quaid-E-Azam
Lahore 3 Tel. 66839
Telex: 44886 UBL PK. Attn: MIRZA BK

Portugal
Livraria Portugal
Rua do Carmo 70-74
1117 Lisboa Codex Tel. 347.49.82/3/4/5

Singapore/Malaysia
Singapour/Malaisie
See "Malaysia/Singapore"
Voir "Malaisie/Singapour"

Spain - Espagne
Mundi-Prensa Libros S.A.
Castello 37, Apartado 1223
Madrid 28001 Tel. (91) 431.33.99
Telex: 49370 MPLI Telefax: (91) 275.39.98

Libreria Internacional AEDOS
Consejo de Ciento 391
08009 -Barcelona Tel. (93) 301-86-15
Telefax: (93) 317-01-41

Sweden - Suède
Fritzes Fackboksföretaget
Box 16356, S 103 27 STH
Regeringsgatan 12
DS Stockholm Tel. (08)23.89.00
Telex: 12387 Telefax: (08)20.50.21
Subscription Agency/Abonnements:

Wennergren-Williams AB
Box 30004
104 25 Stockholm Tel. (08)54.12.00
Telex: 19937 Telefax: (08)50.82.86

Switzerland - Suisse
OECD Publications and Information Centre
4 Simrockstrasse
5300 Bonn (Germany) Tel. (0228)21.60.45
Telex: 8 86300 Bonn
Telefax: (0228)26.11.04

Librairie Payot
6 rue Grenus
1211 Genève 11 Tel. (022)731.89.50
Telex: 28356

Maditec S.A.
Ch. des Palettes 4
1020 Renens/Lausanne Tel. (021)635.08.65
Telefax: (021)635.07.80

United Nations Bookshop/Librairie des Nations-Unies
Palais des Nations
1211 Genève 10
Tel. (022)734.60.11 (ext. 48.72)
Telex: 289696 (Attn: Sales)
Telefax: (022)733.98.79

Taïwan - Formose
Good Faith Worldwide Int'l. Co. Ltd.
9th Floor, No. 118, Sec. 2
Chung Hsiao E. Road
Taipei Tel. 391.7396/391.7397
Telefax: (02) 394.9176

Thailand - Thalande
Suksit Siam Co. Ltd.
1715 Rama IV Road, Samyan
Bangkok 5 Tel. 251.1630

Turkey - Turquie
Kültur Yayinlari Is-Türk Ltd. Sti.
Atatürk Bulvari No. 191/Kat. 21
Kavaklidere/Ankara Tel. 25.07.60
Dolmabahce Cad. No. 29
Besiktas/Istanbul Tel. 160.71.88
Telex: 43482B

United Kingdom - Royaume-Uni
H.M. Stationery Office
Gen. enquiries Tel. (01) 873 0011
Postal orders only:
P.O. Box 276, London SW8 5DT
Personal Callers HMSO Bookshop
49 High Holborn, London WC1V 6HB
Telex: 297138 Telefax: 873.8463
Branches at: Belfast, Birmingham, Bristol, Edinburgh, Manchester

United States - États-Unis
OECD Publications and Information Centre
2001 L Street N.W., Suite 700
Washington, D.C. 20036-4095
Tel. (202)785.6323
Telex: 440245 WASHINGTON D.C.
Telefax: (202)785.0350

Venezuela
Libreria del Este
Avda F. Miranda 52, Aptdo. 60337
Edificio Galipan
Caracas 106
Tel. 951.1705/951.2307/951.1297
Telegram: Libreste Caracas

Yugoslavia - Yougoslavie
Jugoslovenska Knjiga
Knez Mihajlova 2, P.O. Box 36
Beograd Tel. 621.992
Telex: 12466 jk bgd

Orders and inquiries from countries where Distributors have not yet been appointed should be sent to: OECD Publications Service, 2 rue André-Pascal, 75775 Paris Cedex 16.

Les commandes provenant de pays où l'OCDE n'a pas encore désigné de distributeur devraient être adressées à : OCDE, Service des Publications, 2, rue André-Pascal, 75775 Paris Cedex 16.

1/90

OECD PUBLICATIONS, 2, rue André Pascal, 75775 PARIS CEDEX 16
PRINTED IN FRANCE
(61 89 11 1) ISBN 92-64-13324-0 - No. 45099 1990